Intrusion Detection in Wireless Ad-Hoc Networks

Intrusion Detection in Wireless Ad-Hoc Networks

Edited by

Nabendu Chaki · Rituparna Chaki

CRC Press
Taylor & Francis Group
Boca Raton London New York

CRC Press is an imprint of the
Taylor & Francis Group, an **informa** business

CRC Press
Taylor & Francis Group
6000 Broken Sound Parkway NW, Suite 300
Boca Raton, FL 33487-2742

Printed on acid-free paper
Version Date: 20130808

International Standard Book Number-13: 978-1-4665-1565-9 (Hardback)

Library of Congress Cataloging-in-Publication Data

Intrusion detection in wireless ad-hoc networks / editors, Nabendu Chaki and Rituparna Chaki.
pages cm
Includes bibliographical references and index.
ISBN 978-1-4665-1565-9 (alk. paper)
1. Wireless communication systems--Security measures. I. Chaki, Nabendu. II. Chaki, Rituparna.

TK5102.85.I57 2014
005.8--dc23 2013031730

Visit the Taylor & Francis Web site at
http://www.taylorandfrancis.com

and the CRC Press Web site at
http://www.crcpress.com

To our beloved mothers, Hasi Chaki and Tamishra Chatterjee

Nabendu and Rituparna

Contents

Preface

It's a great pleasure to introduce this book on intrusion detection systems (IDSs) for wireless ad-hoc networks. The book aims to ease the job of future researchers working in the field of network security.

This book covers the security aspects for all the basic categories of wireless ad-hoc networks and related application areas, with a focus on IDSs. The categories included are mobile ad-hoc networks (MANETs), wireless mesh networks (WMNs), and wireless sensor networks (WSNs). In the book's eight chapters, the state-of-the-art IDSs for these variants of wireless ad-hoc networks have been reviewed and analyzed. The book also presents advanced topics, such as security in the smart power grid, securing cloud services, and energy-efficient IDSs.

It has been organized in the form of an edited volume with eight chapters, each coauthored by one or more of our senior research scholars and ourselves. We thank and appreciate Novarun Deb, Manali Chakraborty, Debdutta Barman Roy, and Tapalina Bhattasali not only for their chapter contributions, but also for their hard work and innovative suggestions in preparing the manuscript. We especially mention Novarun and Manali for going through the drafts of all the chapters again and again toward improving the quality and content. This has gone a long way toward making this a comprehensive research title in the area of intrusion detection for wireless ad-hoc networks.

We express our sincere thanks to Richard O'Hanley for his continual support and positive influence right from the point of offering us to work on a book on this topic. We are grateful to the publisher for extending us the opportunity to be CRC Press authors. It has been a nice experience to work with Stephanie Morkert and Judith Simon of Taylor & Francis Group during the project and its production process.

Lastly, we must mention and thank every member of our family for their support. We are lucky to have kids like Rikayan and little Nandini, who spared us and sacrificed their valuable time to let us concentrate on the book.

We are sure that all these sacrifices will turn into delight when this book helps budding scholars to explore the area of network security and inspires them to go beyond the covers of this book to craft even better contributions.

Nabendu Chaki
Rituparna Chaki
Kolkata, India

About the Editors

Nabendu Chaki is a senior member of the Institute of Electrical and Electronics Engineers (IEEE) and an associate professor in the Department of Computer Science and Engineering, University of Calcutta, India. Besides editing several volumes for Springer in the Lecture Notes in Computer Science (LNCS) and other series, Nabendu has authored two textbooks and more than 100 refereed research papers in journals and international conferences. His areas of research interests include distributed systems and network security.

Dr. Chaki has also served as a research assistant professor in the Ph.D. program in software engineering in the U.S. Naval Postgraduate School, Monterey, California. He is a visiting faculty member for many universities, including the University of Ca'Foscari, Venice, Italy. Dr. Chaki is a knowledge area editor in mathematical foundation for the SWEBOK project of the IEEE Computer Society. In addiition to being on the editorial board for Springer Verlag and many international journals, he has also served on the committees of more than 50 international conferences.

Rituparna Chaki has been an associate professor in the A.K. Choudhury School of Information Technology, University of Calcutta, India, since June 2013. She joined academia as a faculty member at the West Bengal University of Technology in 2005. Before that, she served under the government of India in maintaining its industrial production database.

Rituparna completed her Ph.D. from Jadavpur University in 2002. She has been associated with organizing many conferences in India and abroad as program chair, organizing chair, or member of the Technical Program Committee. She has published more than 60 research papers in reputed journals and peer-reviewed conference proceedings. Her research interest is primarily in ad-hoc networking and its security. She is a professional member of IEEE and ACM.

Contributors

Tapalina Bhattasali
Department of Computer
Science and Engineering
University of Calcutta
Calcutta, India

Nabendu Chaki
Department of Computer
Science and Engineering
University of Calcutta
Calcutta, India

Rituparna Chaki
A.K. Choudhury School of
Information Technology
University of Calcutta
Calcutta, India

Manali Chakraborty
Department of Computer
Science and Engineering
University of Calcutta
Calcutta, India

Novarun Deb
Department of Computer
Science and Engineering
University of Calcutta
Calcutta, India

Debdutta Barman Roy
Department of Computer
Science and Engineering
Calcutta Institute of Engineering
and Management
Kolkata, India

1

INTRODUCTION

NOVARUN DEB, MANALI CHAKRABORTY, AND NABENDU CHAKI

Contents

1.1 How Security Is Different for a Wireless Ad-Hoc Network (WAHN)?

Traditional wired networks are relatively more secure compared to their wireless counterparts. Conventional networking infrastructure allows traffic through different types of routing devices often laid in a hierarchy. Thus, devices like firewalls and unified threat management (UTM) boxes installed in these routing devices, like switches, gateways, etc., may be highly effective in blocking any intrusion from the outside. Intrusion prevention techniques requiring energy intensive computations may be deployed for securing such a network as there are no power constraints for the proper functioning of these networks. On the other hand, the peer-to- peer multi-hop routing used in wireless ad-hoc networks assumes a completely trusted environment for its basic functioning. This assumption unfortunately remains somewhat impractical, as has been analyzed and described in the chapters of this book. Wireless ad-hoc networks are exposed to a vast array of threats because the wireless medium is inherently insecure. The domain of attacks is transient in nature as are the wireless networks themselves. More importantly, the mobile nodes forming such a network have limited battery life and require periodic replenishment of energy. This prevents the deployment of routing protocols and security solutions that are used in traditional wired networks. Both routing protocols and security solutions have to be redesigned to make them more energy – efficient and less computation intensive. This is why intrusion prevention systems give way to intrusion detection systems.

Intrusion detection systems (IDSs) are well-suited for wireless networks as they retain the ad-hoc and distributed nature of these networks. As a result, the number of packets exchanged is limited and the computational overheads are minimized. Intrusion detection systems are reactive in nature as compared to their proactive counterparts, intrusion prevention systems. They do not ensure a communication environment free from intrusion. Instead, these systems periodically check the network state for any kind of intrusions and raise an alert to the system administrators if any such anomaly is discovered.

1.2 Standards for Wireless Ad-Hoc Networks

Today's wireless ad-hoc networks integrate with multiple wireless systems, like wireless local area network (WLAN), wireless personal area network (WPAN), and wireless metropolitan area network (MAN), to improve the performance of the wireless network and enlarge the coverage range. One of the problems that the IEEE 802.11 system meets during large-scale applications is limited coverage capability. Limited by the transmission power, WLAN can only cover up to 100 m. Access points (APs) can be added to increase the coverage. However, that too adds to the construction cost of the network. As a new networking technology, the wireless ad-hoc network (WAHN) provides a new path to solve these problems. The establishment of WAHN-specific standards is in critical need. With the increasing number of applications of the WAHN, the IEEE 802 standard group started working on related technical standards. During the mid-1980s the IEEE 802.11 working group for WLANs was formed to create a wireless local area network standard. The primary task of IEEE 802.11 was concerned with features such as Ethernet matching speed, handling of seamless roaming, message forwarding, data throughput of 2–11 Mbps, etc. Eventually, it turned out that an even smaller coverage area is needed for higher user densities and the emergent data traffic. The reach of a WPAN is the space around a person or object, extending up to a few meters. Subsequently, the IEEE 802.15 working group was formed to create the WPAN standard. The IEEE 802.15.1 standard [22] proposed by this group is used as the Bluetooth standard. This is the standard for medium-rate WPAN. The low-rate WPANs, IEEE 802.15.4, serve a set of applications with very low cost requirements and with not so stringent needs for data rate and QoS. The low data rate enables IEEE 802.15.4 devices to consume very little power.

The emergence of communication standards such as IEEE 802.15.4 [23, 24] and ZigBee [25, 26], which are targeted at radio frequency (RF) applications requiring low data rate, long battery life, and secure networking, has changed the perception of wireless technologies for sensor networks. With every new wireless standard, several issues are related. While the major concerns include the cost, performance, and quality of the new standard, the lesser issues tend to be energy and security. The major issue of the growth of the wireless sensor network (WSN) applications using the newer standards depends on the availability of integrated chip solutions at an affordable price. This has been witnessed for previous wireless standards-based solutions such as Bluetooth and WLAN. In some applications, interference with other wireless standards like Bluetooth or WLAN is to be addressed. Bluetooth uses channel hopping for passing data. This presents only a momentary state of interference to the wireless sensor network.

1.2.1 Bluetooth

Bluetooth is a packet-based protocol with a master–slave structure [27]. This is a proprietary open wireless technology standard managed by the Bluetooth special interest group (SIG). The SIG has more than 15,000 member companies in the areas of telecommunication, computing, networking, and consumer electronics. It allows the exchanging of data over short distances using the industrial, scientific, and medical (ISM) band from 2400 MHz to 2480 MHz. Bluetooth was originally conceived as a wireless alternative to RS-232 data cables to connect several devices.

Bluetooth uses a radio technology called the frequency-hopping spread spectrum. This splits the data being sent and transmits chunks of it on up to 79 bands of 1 MHz each, allowing for guard bands in between. Using Bluetooth technology, one master may communicate with up to seven slave devices. Such a network is called a piconet. All of the slave devices in a piconet share the master's clock. Packet exchange is based on the basic clock, defined by the master, which ticks at 312.5 μs intervals. The devices can switch roles, by agreement, and the slave can become the master. For example, a mobile phone initiating a connection to another phone has to start the session as master to initiate the connection. However, it may subsequently prefer to be a slave.

Bluetooth technology can also be used to build a larger ad-hoc network connecting more than eight devices. Two or more piconets may be joined to form a scatternet. The piconets in a scatternet may overlap on one or more nodes. However, the constituent piconets may be mutually disjoint as well. In a scatternet, a device may simultaneously play the master role in one piconet and the slave role in another. At any given time, data can be transferred between the master and one of the slave devices. The master chooses the slave device to communicate. Often round-robin scheduling is followed, switching rapidly from one device to another. USB port Bluetooth adapters are commercially referred to as dongles. A Bluetooth dongle has a Bluetooth enumerator and a Bluetooth radio in it. Such devices can link computers with Bluetooth up to a distance of 100 m. Because the devices use a radio link, they do not have to be in visual line of sight of each other. However, a quasi-optical wireless path must be viable. Some of the dongles also include an infrared data association (IrDA) adapter and offer a wide range of services.

1.2.2 Wireless Local Area Network (WLAN)

A wireless local area network (WLAN) links two or more devices using a wireless connection method. WLAN provides a connection to the Internet through an access point. This gives users the mobility to move around within a local coverage area and still be connected to the network. WLANs are based on IEEE 802.11 standards and marketed under the Wi-Fi brand name.

IEEE 802.11 is a set of standards for implementing WLAN communication in the 2.4, 3.6, and 5 GHz frequency bands. The base version of the standard IEEE 802.11-2007 has had several subsequent amendments. The most popular are those defined by the IEEE 802.11b and IEEE 802.11g protocols. These standards provide the basis for wireless network products using the Wi-Fi brand. IEEE 802.11n is a new multistreaming modulation technique. Other standards in the family (c–f, h, and j) are service amendments and extensions or corrections to the previous specifications.

The 802.11b and 802.11g standards use the 2.4 GHz ISM band. As a result, 802.11b and 802.11g devices may suffer interference from microwave ovens, cordless telephones, and Bluetooth devices. Such interference is controlled by using direct-sequence spread spectrum (DSSS) and orthogonal frequency division multiplexing (OFDM) signaling methods. The segment of the radio frequency spectrum used by 802.11 varies in different countries, along with varied licensing requirements. WLAN is designed to exist in parallel with IEEE 802.15.4 in terms of both time and frequency division multiplexing. Using a collision avoidance principle, WLAN listens for a clear RF channel before it talks.

1.2.3 IEEE 802.15.4

The IEEE 802.15.4 is a standard that details the physical layer and media access control (MAC). It is maintained by the IEEE 802.15 working group. As mentioned at the outset of this section, IEEE 802.15.4 is devised for low-rate wireless personal area networks (LR-WPANs). IEEE 802.15.4 forms the basis for the standards that are implemented by different vendors. The list includes ZigBee, ISA 100.11a, WirelessHART, and MiWi specifications. These implementations further enhance the standard by developing the upper layers not specified in 802.15.4.

IEEE standard 802.15.4 intends to offer the fundamental lower network layers of a type of wireless personal area network (WPAN), which focuses on low-cost, low-speed ubiquitous communication between devices. The low data rate enables the IEEE 802.15.4 devices to consume very little power. These features, particularly the low-energy requirement criterion, make IEEE 802.15.4 an obvious candidate for deployment in WSNs.

The basic framework conceives a 10 m communications range with a transfer rate of 250 kbps. Trade-offs are possible to favor even lower energy requirements. Even lower data rates can be considered, with the resulting effect on energy consumption. The primary advantage of 802.15.4 among WPANs is the importance of achieving extremely low manufacturing and operation costs and technological simplicity, without sacrificing flexibility or generality. The important features include carrier sense multiple access with collision avoidance (CSMA/CA) and integrated support for secure communications.

1.2.4 ZigBee

This is a low-data-rate, low-power-consumption, low-cost, wireless networking protocol targeted toward automation and remote control applications. The IEEE 802.15.4 committee started working on a low-data-rate standard in the first decade of the new millennium. Later, the ZigBee Alliance and the IEEE decided to join forces, and ZigBee is the commercial name for this technology. ZigBee-compliant wireless devices are expected to transmit within 10–75 m. This depends on the RF environment and the power output consumption required for a given application. ZigBee devices operate in the unlicensed RF worldwide (2.4 GHz global, 915 MHz Americas, or 868 MHz Europe). The data rate is 250 kbps at 2.4 GHz, 40 kbps at 915 MHz, and 20 kbps at 868 MHz.

ZigBee looks rather like Bluetooth. However, it is simpler. ZigBee has a lower data rate and spends most of its time snoozing. This means that a device on a ZigBee network runs for a long time. The operational range of ZigBee (10–75 m) is also much greater compared to 10 m for Bluetooth. However, ZigBee sits below Bluetooth in terms of data rate. The data rate of ZigBee is 250 kbps at 2.4 GHz, 40 kbps at 915 MHz, and 20 kbps at 868 MHz, whereas that of Bluetooth is 1 Mbps. The ability of IEEE 802.15.4 or a ZigBee-based network to perform automatic retries helps us to overcome interference from Bluetooth devices.

IEEE 802.11s, 802.15, 802.16, and 802.20 are some other standards that have specifications for a rather recent variation of ad-hoc networks called wireless mesh networks (WMNs). The typical functionalities included in these standards are access control, routing, congestion control, mobility, handoff support, and authentication for security.

1.2.5 IEEE 802.11s

The IEEE 802.11s is an IEEE 802.11 amendment for mesh networking to be used for both static topologies and ad-hoc networks. It extends the IEEE 802.11 MAC standard by defining an architecture and protocol that supports both broadcast/multicast and unicast delivery. IEEE 802.11s is inherently built upon 802.11a, 802.11b, 802.11g, or 802.11n carrying the actual traffic. The appropriate routing protocol is used, depending on the physical topology of the network. The hybrid wireless mesh protocol (HWMP), inspired by a blend of the ad-hoc on-demand distance vector (AODV) routing and tree-based routing, is used for 802.11s as a default.

In a cellular network, there will often be a handoff from one base station to another. This could be from IEEE 802.11 as well as from networks following global system for mobile communication (GSMC), Bluetooth, process control system (PCS), and other cordless protocols. IEEE 802.11s ensure such handoffs between nodes, obeying 802.11s and otherwise. IEEE 802.11s also includes mechanisms to provide deterministic network access, a framework for congestion control and power saving.

IEEE 802.11s activities started as a study group of IEEE 802.11 in 2003, which subsequently became a task group in 2004. The first draft standard for IEEE 802.11s was accepted after a unanimous confirmation vote in 2006. Since then, the draft has evolved through informal comment resolution. In June 2011 the fifth recirculation sponsor ballot on TGs Draft 12.0 was closed. The draft met with 97.2% approval.

1.2.5.1 Prerelease Implementation of IEEE 802.11s IEEE 802.11s aims to get over the operational limitations of the traditional AP. The service flow can be forwarded to adjacent APs for multihop transmission. In this way, the WMN is provided with higher reliability, better scalability, and lower investment cost. Therefore, in the new WLAN structure, the APs form the WMN backbone network of the WLAN automatically.

Although the final standard for 802.11s is yet to be released, this has been accepted as the future standard for wireless mesh networks. This is evident from the current release of different versions of operating systems (OSs). A reference implementation of the 802.11s draft is available as part of the MAC layer in the Linux kernel, starting with version 2.6.26. The Linux community, with its many diverse distributions, provides a heterogeneous testing ground for protocols like HWMP. In FreeBSD operating systems too, the 802.11s draft is supported starting with FreeBSD 8.0.

1.2.6 IEEE 802.11 and Security in WMN

Mesh networking often involves network access by previously unknown parties, especially when a transient visitor population is being served. The IEEE 802.11u standard will be required by most mesh networks to authenticate these users without preregistration or any prior offline communication. Prestandard captive portal approaches are also common.

There are no defined roles of nodes in a mesh. Therefore, the security protocols for mesh must be true peer-to-peer protocols where either side can initiate to the other or both sides can initiate simultaneously. IEEE 802.11s defines a secure password-based authentication and key establishment protocol between the peers. This is called simultaneous authentication of equals (SAE). SAE is based on a zero knowledge proof and is resistant to active attack, passive attack, and dictionary attack.

When peers discover each other, they take part in an SAE exchange. If SAE completes successfully, each peer would know that the other party possesses the mesh password. Consequently, the two peers establish a cryptographically strong key. This key is exchanged like a secret key with the authenticated mesh peering exchange (AMPE) to establish a secure peering and derive a session key to protect mesh traffic, including routing traffic.

Clients wishing to send and receive packets locally without routing them for others can use simpler approaches, such as 802.11u authentication. This does not require preauthentication. This is not part of the specification of the mesh network itself.

1.2.7 IEEE 802.15 Standards

IEEE 802.15 standard family is developed for the wireless WPAN. These standards define the physical and MAC layers of WPAN. At present, IEEE 802.15.1, 802.15.2, and 802.15.3 are not designed to support the WMN structure. However, these can support the Bluetooth piconet structure, and scatternet is an important element of the WMN. The IEEE 802.15.4 standard is meant for the application devices that ask for low data rates and long battery life. It provides the WPAN with an integrated, energy-efficient network solution.

IEEE 802.15.5, developed for the MAC layer of WMN, is still under research. It follows some basic ideas used in IEEE 802.15.1 through 802.15.4. However, IEEE 802.15.5 fully supports the mesh structure, with no need of support from ZigBee or the internet protocol (IP) route. In the 802.15.5 standard, the mesh network is defined as a PAN with two networking modes. These are full- and partial-mesh architectures. In the full-mesh version, any node can be connected to other nodes directly. In the partial-mesh topology, some nodes may be allowed to connect directly to other nodes. The other nodes only connect to the nodes with high data exchange rates. The major issues addressed by IEEE 802.15.5 include the following:

- Collision sense beacon scheduling
- Security issues
- Energy efficiency
- Support of mesh nodes and the mobility of mesh PAN
- Routing algorithm

1.2.8 IEEE 802.16 Standards

IEEE 802.11 networks fulfill the requirements for data services in a local area. IEEE 802.16 aims to serve the broadband wireless access in metropolitan area networks. WiMAX is the commercialization of the maturing IEEE 802.16 standard. The original 802.16 standard operates in the 10–66 GHz frequency band and requires line-of-sight towers.

The 802.16a extension uses a lower frequency of 2–11 GHz. This frees up the line-of-sight connection requirement. The 802.16a standard is able to connect more customers to a single tower and substantially reduce service costs. An extension of IEEE 802.16, called IEEE

802.16e, is being developed to allow users to connect to the Internet while moving at vehicular speeds.

IEEE 802.16 is primarily meant for peer to multipoint (P2MP) technology. With the development of the WMN, the IEEE 802.16 standard group introduces the mesh structure into the IEEE 802.16d/e standard that has been put forward recently. WMN uses broadcasting or multicasting to transmit between nodes. Therefore, the link disconnection in WMN is less crucial than in a typical P2MP system. With the increase in the number of nodes, the IEEE 802.16 mesh network can be more robust and have a wider coverage.

The IEEE 802.16 standard supports two different physical layers and supports adaptive modulation and coding. Therefore, the link rate changes with different channel conditions. In WMN-based WMAN, users can communicate with each other directly or indirectly without using any base station (BS). The signals would be transmitted in the hop-by-hop mode. This adaptation on WMAN can increase the system coverage. When new subscribers are added, the network can change its topology dynamically, avoiding the increase of base stations.

IEEE 802.16d supports overall and efficient scheduling of resources based on the time division multiple access (TDMA) mode. IEEE 802.16e standards are developed to further support user mobility. The IEEE 802.16e system supports both local and regional mobility, roaming, and handoff, and provides a rate up to 150 km/h in moving environments. IEEE 802.16 mesh in the current standard draft has several limitations:

- The 802.16 mesh has limited scalability. The mesh can only support around 100 subscribers due to centralized scheduling message structures.
- The 802.16 mesh is based on a connectionless MAC. Thus, QoS of real-time services is difficult to guarantee.
- It is assumed that there is no interference between nodes that are two hops away. Thus, the 802.16 mesh suffers from the hidden terminal problem.

A group within 802.16, the Mesh Ad-Hoc Committee, is investigating ways to improve the performances of mesh networking. This includes study of peer-to-peer (P2P) data transmission support and signal obstacle traversal.

1.2.9 IEEE 802.20 Standards

The IEEE 802.20 network is a pure IP-based system. The IEEE 802.20 standard working group aims to develop standards for mobile subscribers. IEEE 802.20 has the following two major goals:

- Combining the advantages of the high data rates of the fixed wireless access network and the high mobility of the cellular network to resolve the conflicts between the fixed wireless APs and the increasing demands on high-rate mobile services
- Specification of a global standard for the mobile broadband wireless access network that supports diversified IP services

IEEE 802.20 standards are to support high QoS, high frequency efficiency, and reliable high-speed wireless data transmission on the 3 GHz band. Therefore, IEEE 802.20 is expected to provide users, moving at high speed and working on a broad-bandwidth, with data transmission speeds of 1 Mbps or higher. This would enable applications like video conferencing, or video-medic, even on a high-speed train. IEEE 802.20 improves the current IEEE 802.16 performance of low mobility with high data rate, and high mobility with low data rate of the cellular network. It supports WMN in either indoor or outdoor environments. In the IEEE 802.20 mesh architecture, the mobile nodes can communicate with each other. This would support improving the routing performance for the mobile network. Besides, this facilitates fast access to the backbone network and provides mobile users with quick and accurate services.

1.3 Protocols for Wireless Ad-Hoc Networks

Wireless networks are relatively new compared to wired networks. Two distinct approaches for developing protocols of wireless networks exist: modify the existing protocols of wired networks or develop from scratch. Most of the layer specific protocols are discussed in this section.

1.3.1 MAC Protocols

Nodes in mobile ad-hoc networks (also known as mobile stations) are attached to a transmitter/receiver that communicates via a wireless channel shared by other nodes. Transmission from any mobile station is received by other mobile stations in the neighborhood. The wireless channel is thus shared among multiple neighboring nodes. If multiple nodes try to communicate over the wireless medium simultaneously, then a collision occurs. Hence, it becomes critical to decide which mobile node has access to the wireless channel at a particular instant in time.

Wireless medium access control (WMAC) protocols define rules for orderly access to the shared wireless medium. Sharing of the wireless link should be fair and efficient in terms of bandwidth. WMAC mechanisms need to avoid packet collisions due to interference. There are two different types of WMAC protocols:

1. *Contention protocols* resolve a collision after it occurs or try to avoid it. These protocols execute a collision resolution protocol after each collision. Examples include carrier sense multiple access with collision avoidance (CSMA/CA), multiple access with collision avoidance (MACA), floor acquisition multiple access (FAMA), multiple access with collision avoidance for wireless (MACAW), etc.
2. *Conflict-free protocols* ensure that a collision can never occur. Examples of these protocols include frequency division multiple access (FDMA), TDMA, and code division multiple access (CDMA).

Mobility and energy constraints of nodes in wireless ad-hoc networks play a vital role in deciding the mechanisms for accessing the wireless channel. All these characteristics make the design of WMAC schemes more challenging than that for the wired networks. There are three important issues related to the design of access control protocols in the wireless medium:

- *Half-duplex operation.* In the wireless medium, it is not easy to receive data while the transmitter is sending data. This is because when a node transmits signals, a large fraction of the signal energy leaks into the receiver path. The transmitting and receiving power levels can differ by orders of magnitude. The leakage signal typically has much higher power than the received signal, which implies that it is impossible to detect a received signal while data are being transmitted over the wireless medium. Collision detection is not possible while sending data, and CSMA/CD (Ethernet MAC) cannot be used as it is. As collision cannot be detected by the sender, all proposed protocols attempt to minimize the probability of collision, and hence focus on collision avoidance.

- *Time-varying channel.* Radio signal propagation is governed by three different mechanisms: *reflection*, *diffraction*, and *scattering*. The signal received by a node is the juxtaposition of time-shifted and attenuated versions of the transmitted signal. This time-varying channel phenomenon is also known as multipath propagation. The rate of variation of the channel is determined by the coherence time of the channel. Coherence time is defined as time within which the received signal strength changes by 3 dB. When a node's received signal strength drops below a certain threshold, the node is said to be in fade. Handshaking is a widely used strategy to ensure the link quality is good enough for data communication. A successful handshake between a sender and a receiver (small message) indicates a good communication link.
- *Burst channel errors.* As a consequence of time-varying channels and varying signal strengths, errors are introduced in the transmission. For wired networks the bit error rate (BER) is typically 10^{-6}; i.e., the probability of packet error is small. Errors are mainly due to random noise. For wireless networks, the BER is as high as 10^{-3}. Here the errors are due to nodes being in fade. As a result, errors occur in long bursts. Packet loss due to burst errors can be mitigated using the following techniques:

  1. Smaller packets
  2. Forward error correcting codes
  3. Retransmissions (ACKs)

- *Location-dependent carrier sensing.* Wireless signals decay with the square of distance in free space. This implies that carrier sensing becomes a function of the position of the receiver relative to the transmitter. In a wireless medium, due to multipath propagation, the signal strength decays according to a power law with distance; i.e., only nodes within a specific radius of the transmitter can detect the carrier signal on the channel. Location-dependent carrier sensing results in three types of nodes that protocols need to deal with:

1. Hidden nodes. Even if the medium is free near the transmitter, it may not be free near the intended receiver.
2. Exposed nodes. Even if the medium is busy near the transmitter, it may be free near the intended receiver.
3. Capture. Capture occurs when a receiver can cleanly receive a transmission from one of two simultaneous transmissions.

All the above factors necessitate reengineering MAC protocols for the wireless ad-hoc environment. The following sections present some WMAC schemes that may bc applied to the wireless ad-hoc environment [28]. All of them are modifications of the standard IEEE 802.11 distributed coordination function (DCF).

1.3.1.1 IEEE 802.11 DCF CSMA/CA forms the basis of a distributed MAC-based local assessment of the channel status. Before transmitting a frame, if the channel is found to be busy, then the MAC waits for the channel to become idle and waits for a DCF interframe space (DIFS) amount of time. If the channel remains idle during this interval, then it is followed by a binary exponential backoff process that sets a backoff counter (BC) to a random value within a contention window (CW). The BC is decremented with each slot time interval and the frame is transmitted when the BC reaches zero. If a frame arrives when the MAC is in DIFS deference or random backoff, it is transmitted only when the random backoff finishes successfully. If the medium is idle for more than the DIFS interval and no backoff is ongoing, then the frame is transmitted immediately.

The BC is set to a random integer from a uniform distribution over the closed interval [0, CW]. The CW size is initially CW_{min} and increases when a transmission fails. After an unsuccessful transmission attempt, another backoff is performed using a new CW value given by

$$CW_{new} = [2 \times (CW_{old} + 1) - 1]$$

This reduces the collision probability in case there are multiple stations attempting to access the channel. After each successful transmission, the CW value is reset to CW_{min}.

MACA is DCF with the request to send/clear to send (RTS/CTS) mechanism and is somewhat a kind of virtual carrier sensing. The

RTS/CTS mechanism reduces the number of collisions and also solves the hidden and exposed terminal problems.

1.3.1.2 IEEE 802.11b—Enhanced DCF (EDCF) EDCF is deployed as a contention-based medium access mechanism. It differs from DCF in that eight levels of user priorities can be applied to the nodes. A wireless node with higher priority is assigned shorter CW_{min} and CW_{max}, which increases the chances of high-priority flows being transmitted before the lower-priority ones. Also, high-priority stations have longer IFS than low-priority ones, which have shorter IFS. The IFS is called arbitrary IFS (AIFS). EDCF is designed to provide good prioritization and distributed channel access for frames with different levels of priority. It can support real-time applications with voice and video traffic with a reasonable quality of service in certain environments. However, like DCF, it performs poorly when the traffic load is high due to frequent collisions and wasted idle time. Furthermore, EDCF suffers from low-priority traffic starvation, especially at high load conditions, which impairs its fairness.

1.3.1.3 Adaptive EDCF (AEDCF) Differentiated access provided by EDCF provides better QoS to high-priority classes while minimizing service to low-priority traffic. This scheme is successful in providing high QoS to real-time traffic but fails to capture other wireless ad-hoc network conditions, such as collision rate and network load. This drawback is overcome by AEDCF, which updates the CW parameter as a function of the network conditions. This further improves the QoS of services provided over the network.

After a successful transmission, the CW value is updated gradually rather than resetting it to CW_{min}. This serves the purpose of avoiding busy collisions. Similarly, the backoff process after a collision is random but no longer a binary exponential process. In other words, the CW value is not doubled but increased with a persistence factor with the purpose of increasing the CW of high-priority traffic slower than that of low-priority traffic.

The gradual update of the CW parameter considers the average collision rate at each station, which is updated periodically. The calculation is such that flows with a high collision rate have a better chance to transmit during the next contention slot. AEDCF decreases the collision rate between stations with the same priority and decreases the access latency as well. Thus, AEDCF is an extension of EDCF that is more adaptive to the network conditions. The problem of AEDCF is that the performance of background low-priority traffic flows degrades at high load, as it has a much larger average CW size than high-priority traffic. This increases the waiting time and impairs channel utilization.

1.3.1.4 Adaptive Fair EDCF (AFEDCF) The performance of distributed contention-based approaches is impaired by packet collisions and wasted idle slots resulting from the backoff process in each contention cycle. The ideal situation demands that successive packet transmissions are not interspersed with these impairments. AFEDCF tries to do exactly that.

The AFEDCF mechanism increases the CW value not only during collisions but also when the medium is sensed busy during deferral periods. The backoff timer is decreased when the medium is sensed idle. Decrementing the counter occurs in two phases: linear decrease and fast decrease. In the *linear decrease* phase, the backoff counter (BC) is decremented by a fixed amount during each idle time slot, whereas in the *fast decrease* phase, the counter decrements exponentially.

The threshold value identifying the boundary between these two stages is variable. Depending on the traffic load, the threshold should be increased during low contention periods and decreased during high contention periods. When a collision occurs or a wireless node is waiting for the channel to be idle, it doubles the CW value, randomly chooses a new backoff time, and reduces the backoff threshold to make the fast decrease phase shorter. When a packet is transmitted successfully, the wireless node resets the CW value to CW_{min}, randomly chooses the backoff time, and increases the backoff threshold to make the fast decrease phase longer. AFEDCF does not use the adaptive CW update mechanism like AEDCF does.

AFEDCF exhibits high performance by achieving high throughput and fairness even during high load conditions. Fairness is achieved because CW sizes of wireless nodes reach their maximum values rapidly during high traffic load conditions. This implies that the nodes will be transmitting almost all the time at the same contention window. A varying backoff threshold mechanism results in reduced wasted idle time slots and adaptation to collision rate.

1.3.2 Network Layer Protocols

The network layer is responsible for *routing* variable-length data sequences from a source to a destination host via one or more networks. Maintaining the quality of service (QoS) is essential while performing this functionality. The main function of routing packets is done by the Internet protocol (IP). All other QoS functionalities are provided by other protocols, such as address resolution protocol (ARP), reverse address resolution protocol/dynamic host configuration protocol (RARP/DHCP), Internet control message protocol (ICMP), Internet group management protocol (IGMP), etc. These QoS protocols supplement the functioning of IP. Functions of the network layer include:

1. Connection model: connectionless communication. IP is a connectionless protocol following which a packet can travel from the source to the destination node without the receiver having to return an acknowledgment. Another major characteristic of connectionless communication is that successive packets may be routed through completely or partially disjoint routes.
2. Host addressing. Since the network layer is responsible for host-to-host delivery, every host within the network must have an address identifier associated with it. This address is used to identify a host uniquely within the network and follows a hierarchy. On the Internet, these addresses are known as IP addresses.
3. Message forwarding. Wide area communications are achieved by interactions across different types of networks as well as between subnetworks within a network. Specialized hosts, called gateways and routers, are used to forward packets across wide area networks.

The entire scenario changes when the network becomes wireless ad-hoc in nature and the nodes execute mobile applications. The network layer must ensure that messages for a particular host reach it even when it moves from one location to another. IP was not designed with this feature in mind.

Figure 1.1 clearly illustrates the problem that arises when a node moves within the network. The network node 2.0.0.4 moves from network B to network C. However, its IP does not change. As a result,

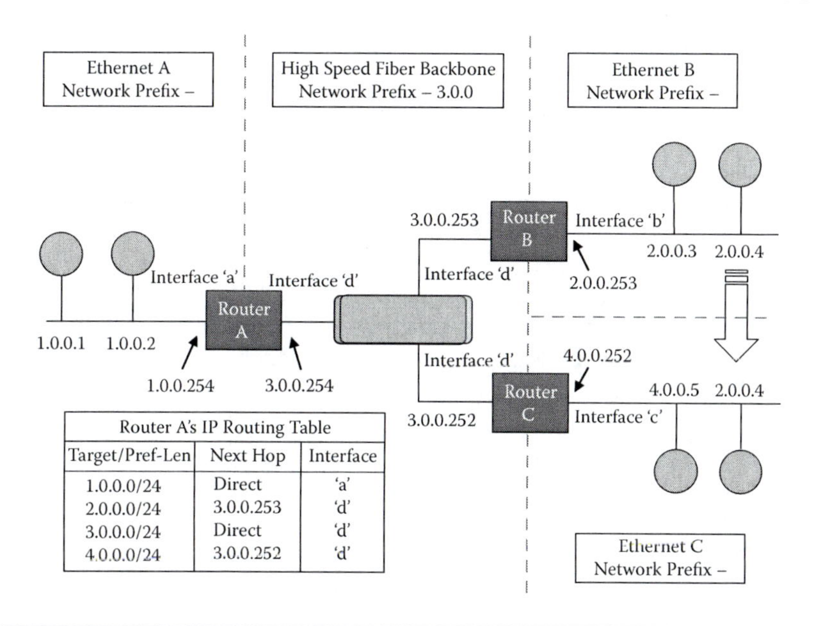

Router A's IP Routing Table		
Target/Pref-Len	Next Hop	Interface
1.0.0.0/24	Direct	'a'
2.0.0.0/24	3.0.0.253	'd'
3.0.0.0/24	Direct	'd'
4.0.0.0/24	3.0.0.252	'd'

Figure 1.1 Node movement from one subnet to another.

when router A wants to send a packet to this node, it finds that the next hop to forward the packet is 3.0.0.253, that is, router B. Thus, the packet fails to reach the desired destination node. This is where mobile IP becomes significant.

1.3.2.1 Mobile IPv4 Alternate solutions can be thought of for provisioning the mobility of network nodes in a wireless ad-hoc network. However, these approaches do not prove to be feasible due to the overhead involved. Some of these approaches are as follows:

- *Changing the IP address.* This is the most obvious solution with the most obvious drawback. Changing the IP address whenever a node changes its point of contact makes it impossible for the node to maintain transport and higher-level connections between the source and destination. The Transmission Control Protocol (TCP) maintains connections that are indexed by a quadruplet: <source IP, source port, destination IP, destination Port>. Thus, if the IP address is changed, the TCP connection will break. Reestablishing these connections does not solve the problem, as the node may keep changing its connectivity with the network.

- *Applying link layer solutions.* Link layer solutions to node mobility exist, such as cellular digital packet data (CDPD). However, such solutions provide node mobility only in the context of a single type of medium and within a limited geographic area. It is quite obvious that these link layer solutions cannot be applied to the vast Internet domain, which is an integration of different types of networks.
- *Host-specific routes.* There can be numerous nodes in a wireless ad-hoc network. Maintaining host-specific routes incurs a huge overhead. Every time a node moves from one link to another, all routes associated with that node have to be updated. Furthermore, host-specific routes have to be propagated throughout much of the Internet routing fabric. Minimally, these routes must be propagated between a mobile node's home link and foreign link. Host-specific routing has severe scalability, robustness, and security issues. Thus, host-specific routing is not a feasible solution to node mobility in the Internet.

Mobile IP is the network layer solution to node mobility in the Internet. It accomplishes its task by setting up the routing tables in appropriate nodes, such that IP packets can be sent to mobile nodes not connected to their home link. It can be considered a routing protocol, which has a very specialized purpose of allowing IP packets to be routed to mobile nodes, which could potentially change their location very rapidly. Mobile IP is unique in its ability to accommodate heterogeneous mobility in addition to homogeneous mobility. It solves the primary problem of routing IP packets to mobile nodes, which is a first step in providing mobility on the Internet. A complete mobility solution would involve enhancements to other layers of the protocol stack.

Mobility agents (*home agents* and *foreign agents*) advertise their presence by periodically multicasting or broadcasting *agent advertisement* messages. Mobile nodes listen to these advertisements and examine their contents to determine whether they are connected to their home link or to a foreign link. When a mobile node detects that it is in its home network, it operates without mobility services. When the node detects that it has moved to a foreign network, it obtains a care-of-address (COA). A foreign agent care-of address can be read from one of the fields within the foreign agent's agent advertisement.

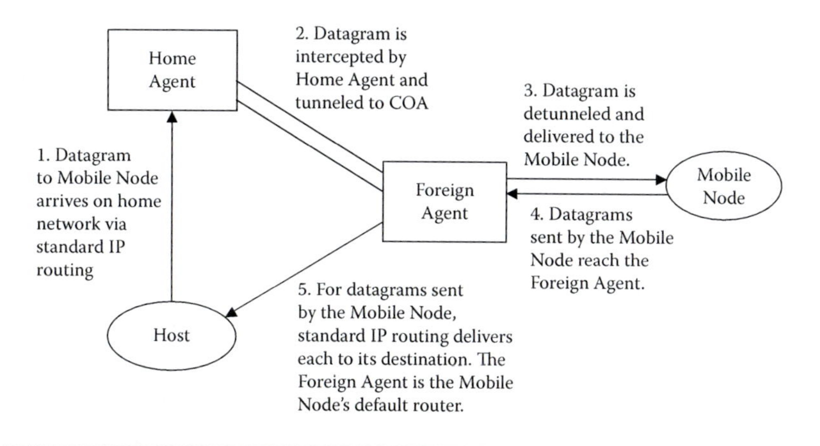

Figure 1.2 Mobile IPv4 routing mechanism.

A collocated care-of address must be acquired by some assignment procedure, such as dynamic host configuration protocol (DHCP), the point-to-point protocol's IP control protocol (IPCP), or manual configuration. The mobile node then registers its newly acquired COA with its home agent through exchange of *registration* messages, possibly through a foreign agent.

Figure 1.2 illustrates how datagrams are routed using mobile IPv4. The home agent intercepts datagrams sent to the mobile node's home address, and *tunnels* them to the COA that the mobile node previously registered. At the COA (either the foreign agent or one of the interfaces of the mobile node itself) the original packet is extracted from the tunnel and then delivered to the mobile node. In the reverse direction, datagrams sent by the mobile node are generally delivered to their destination using standard IP routing mechanisms, not necessarily passing through the home agent. The foreign agent serves as a *default router* for all packets generated by a visiting node. When a mobile node returns to its home network, it deregisters itself with its home agent through the exchange of registration messages.

1.3.2.2 Mobile IPv6 A mobile node is always expected to be addressable at its home address, irrespective of whether it is attached to its home link. While a mobile node is at home, packets addressed to its home address are routed to the mobile node's home link using standard Internet routing mechanisms. While a mobile node is attached to some foreign link, it is also addressable at one or more care-of-addresses (COAs). As long as the mobile node stays in this location, packets addressed to this COA will be routed to the mobile node. The mobile node may also accept packets from several COAs, such as when it is moving but still reachable at the previous link. There are two possible modes for communication between the mobile node and a corresponding node.

1. *Bidirectional tunneling.* This mode does not require mobile IPv6 support from the correspondent node and is available even if the mobile node has not registered its current binding with the correspondent node. Packets from the correspondent node are routed to the home agent and then tunneled to the mobile node.

   Packets to the correspondent node are tunneled from the mobile node to the home agent (*reverse tunneled*) and then routed normally from the home network to the correspondent node. In this mode, the home agent uses *proxy neighbor discovery* to intercept any IPv6 packets addressed to the mobile node's home address on the home link. Each intercepted packet is tunneled to the mobile node's primary COA. This tunneling is performed using *IPv6 encapsulation.* The packet headers for bidirectional tunneling are illustrated in Figure 1.3.
2. *Route optimization.* This mode requires the mobile node to register its current binding at the correspondent node. Packets from the correspondent node can be routed directly to the COA of the mobile node. When sending a packet to any IPv6 destination, the correspondent node checks its *cached bindings* for the packet's destination address. If a cached binding

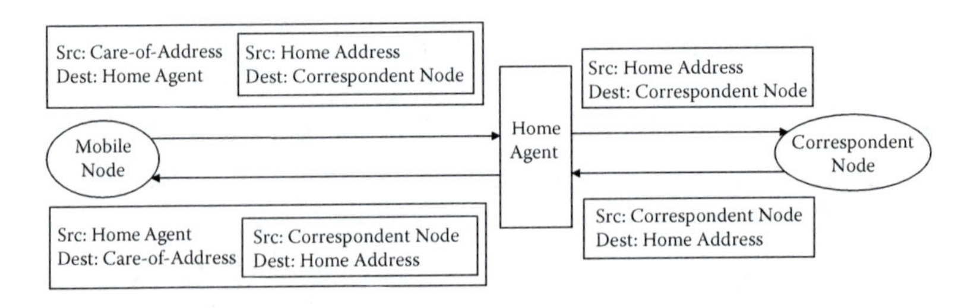

Figure 1.3 Packet structure during bidirectional tunneling.

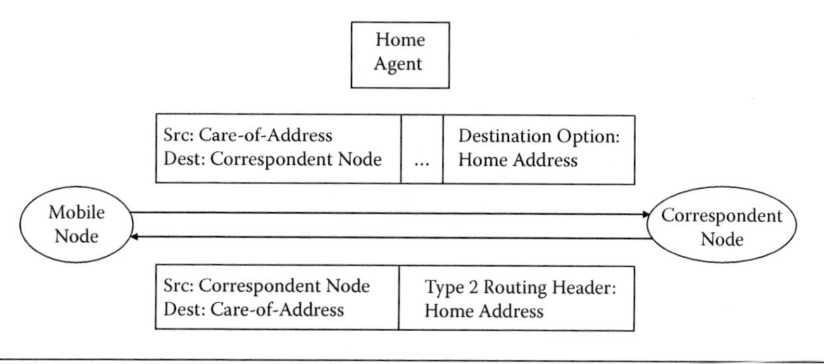

Figure 1.4 Packet structure during route optimization.

is found, the node uses a new type of IPv6 routing header to route the packet to the mobile node by way of the COA indicated in this binding.

Routing packets directly to the mobile node's COA allows the shortest communication path to be used. It also eliminates congestion at the mobile node's home agent and home link. In addition, the impact of any possible failure of the home agent or networks on the path to or from it is reduced. Figure 1.4 illustrates the packet header structures for route optimization.

When routing packets directly to the mobile node, the correspondent node sets the destination address in the IPv6 header to the COA of the mobile node. A new type of IPv6 header is also added to the packet to carry the desired home address. Similarly, the mobile node sets the source address in the packet's IPv6 header to its current COA. The mobile node adds an IPv6 *"home address" destination option* to carry its home address. The inclusion of home addresses in these packets makes the use of COAs transparent above the network layer.

1.3.2.3 Comparison between MIPv4 and MIPv6 Unlike MIPv4, there is no need to deploy special routers like foreign agents in MIPv6. MIPv6 operates in any location without any special support required from the local router. Support for route optimization is a fundamental part of the MIPv6 protocol, rather than a nonstandard set of extensions. MIPv6 route optimization can operate securely even without prearranged security associations. It is expected that route optimization can be deployed on a global scale between all mobile nodes and correspondent nodes. The MIPv6 neighbor unreachability detection assures symmetric reachability between the mobile node and its default router in the current location. Most packets sent to a mobile node while away from home in MIPv6 are sent using an IPv6 routing header rather than IP encapsulation, reducing the amount of resulting overhead compared to MIPv4. MIPv6 is decoupled from any particular link layer, as it uses IPv6 neighbor discovery instead of ARP. This also improves the robustness of the protocol. The dynamic home agent address discovery mechanism in MIPv6 returns a single reply to the mobile node. The directed broadcast approach used in IPv4 returns separate replies from each home agent.

1.3.3 Transport Layer Protocols

The transport layer for wired networks is governed by two important standardized protocols: Transmission control protocol (TCP) and user datagram protocol (UDP). TCP is a connection-oriented protocol and resembles a telephonic communication. Data (voices) are exchanged between the source and destination only after a call connection is established. UDP, on the other hand, is a connectionless protocol and resembles the post office communication. Data (letters) may reach the receiver in any random order and via different traffic routes.

The TCP/IP protocol stack is used globally for wired communication. However, TCP does not perform well when it is used in wireless ad-hoc networks (WAHNs) because of the following:

- *Misinterpretation of packet loss.* The unreliable nature of the wireless medium, collisions, interference, and fading properties of wireless signals results in greater packet loss than in wired networks. TCP is unable to interpret this and act accordingly.
- *Frequent path breaks.* The mobile nature of nodes forming a wireless ad-hoc network results in dynamic changes in network topology. This causes the wireless links between nodes to change frequently. Connectivity between nodes, and hence routes from source to destination, may be recomputed quite often.
- *Effect of path length.* Underwater acoustic network (UAN) is a type of wireless ad-hoc network that is deployed in an aquatic environment. A string topology is an essential component of

UANs that helps improve network performance. However, this particular type of topology results in increased path lengths between sender and receiver. TCP throughput degrades rapidly with an increase in path length, and hence is not suitable for such wireless ad-hoc networks.

- *Misinterpretation of congestion window.* Congestion control mechanisms involved in wired networks get invoked during heavy traffic scenarios. For wireless ad-hoc networks, congestion control mechanisms are required when a network gets partitioned. Network partitioning occurs frequently in wireless ad-hoc networks, and this increases the recovery time objective (RTO) of the network.
- *Asymmetric link behavior.* The radio channel in wireless networks has different properties, such as location-dependent contention, environmental effects on propagation, and directional properties leading to asymmetric links. This affects the performance of TCP.

All of the above factors necessitate rediscovering transport layer protocols for the wireless ad-hoc environment. Transport layer protocols can be engineered under two different architectures. The first type of classification is based on the layer architecture of the open systems interconnection (OSI) stack. There are two categories of transport layer protocols:

1. *Cross-layer solutions.* Cross-layer solutions are dependent on the interaction between any two layers of the OSI stack. The motivation for this classification results from the fact that providing lower-layer information to the upper layers should improve the upper layer performance. Depending on which two layers of the OSI stack are communicating, transport layer protocols can be further classified as:
   a. TCP and network
   b. TCP and link
   c. TCP and physical
   d. Network and physical
2. *Single-layered solutions.* These solutions rely on adapting a layer of OSI stack in isolation that is independent of any other layer. Three such layers have been identified that when

modified give improved performance in wireless ad-hoc networks. These are:

a. TCP layer
b. Network layer
c. Link layer

The classification depicted in Figure 1.5 shows various transport layer protocols. In the second classification transport layer solutions are divided based on the engineering or design approach (Figure 1.6). There are two standard approaches that divide transport layer solutions for wireless links into two categories. These are as follows:

1. *TCP over ad-hoc networks.* These solutions are obtained by tweaking the existing TCP. Redesigning or reengineering TCP implies making improvements to the existing protocol so that it can be applied over wireless ad-hoc network solutions for achieving a better throughput.

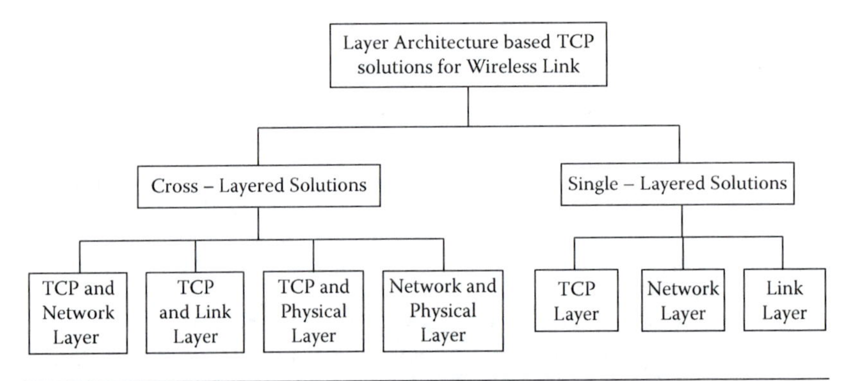

Figure 1.5 Classification of transport layer protocol based on layered architecture.

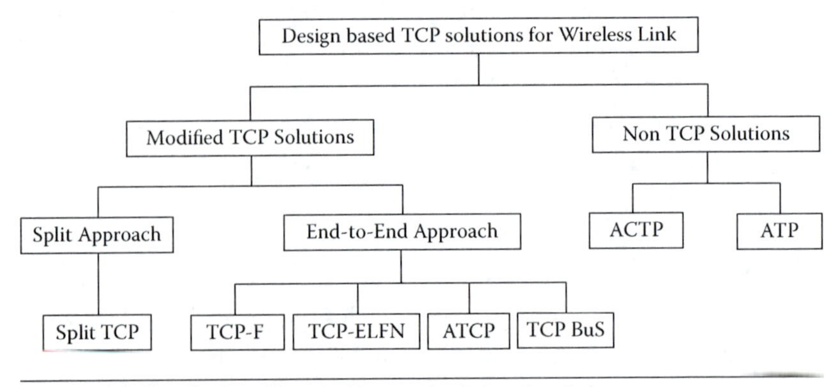

Figure 1.6 Classification of transport layer protocols based on design.

2. *Non-TCPs.* These solutions take a completely different approach. Non-TCP transport layer solutions come up with the idea of developing entirely new protocols, specific to the needs of wireless ad-hoc networks.

1.3.3.1 Split Approach *Split TCP* is a modified TCP solution that splits the transport layer objectives into congestion control and reliability. This is effectively implemented by splitting the TCP connection into two parts. One TCP connection exists between the sender and base station, and another between the base station and the receiver. The TCP sender is completely hidden from the wireless ad-hoc network by ceasing the TCP connection at a base station and using a partition-reliable connection between base station and destination host. The partition connection can utilize selective or negative acknowledgment or some specific protocol adjusted to perform well over the wireless link. Solutions based on this mechanism are I-TCP [7], M-TCP [8], etc.

1.3.3.2 End-to-End Approach

TCP-F. TCP with feedback manages route failures in ad-hoc networks [9]. TCP-F relies on the network layer at intermediate nodes to uncover the path breakdown due to the movement of downstream neighbors with the route. TCP-F places the TCP sender in one of two states: active state and snooze state. In the *active state*, the standard TCP performance is grasped by the TCP sender.

When an intermediate node reveals a link breakdown, it dispatches a route failure notification (RFN) packet to the sender and reports this experience. After getting the RFN, the sender joins the *snooze state*, stops sending more packets, and immobilizes the variable values, such as retransmission timer and congestion window size. The sender waits in the snooze state until the intermediate node observes a reestablishment of the path through a route reestablishment notification (RRN) packet. The sender regains the active state. The link breakage is first identified at the intermediate node. The TCP sender cannot identify it until a special RFN packet appears from the failure point. Similarly, the detection of restoration depends on the special RRN packets from some intermediate nodes. The RFN and RRN packets are relayed to the sender by TCP.

TCP-ELFN. TCP-ELFN [10] is based on explicit link failure notification technique, like TCP-F, but this is an interface between TCP and the routing protocol. The interface plans to update the TCP agent on route failures when they arise. ELFN is based upon the Dynamic Source Routing (DSR) protocol. To execute an ELFN message, DSR route error messages are adapted to carry a payload. As a TCP sender gets an ELFN, it stops its retransmission timers and enters a *"stand-by" mode*, which is similar to the snooze state of TCP-F. In contrast with TCP-F, link breakage information in ELFN is carried with adapted route error messages, which are transmitted under the control of the routing protocol. The link failure information is moved up to the transport layer only at the TCP sender.

ATCP. Ad-hoc TCP [11] utilizes network layer feedback. It depends on the network layer to create correct *ICMP host unreachable messages* and circulate them to the TCP sender. ATCP attempts to deal with the high bit error rate. The TCP sender can be put into a persist state, congestion control state, or retransmit state. ATCP introduces a thin layer between TCP and IP. This layer is on the lookout for explicit congestion notification (ECN) messages and ICMP "Destination Unreachable" messages; it modifies the network state information and then establishes the TCP sender in the appropriate state. After getting a "Destination Unreachable" message, the sender goes into the *persist state*. The TCP at the sender is frozen, and no packet is sent until a new route is established.

The sender does not evoke congestion control, as only ECN messages are used to report to the sender that network congestion has occurred along the route being used. Only on acceptance of an ECN message does the sender enter a *congestion control state* and congestion control mechanisms are evoked without waiting for a time-out event. If a packet loss still follows and the ECN flag is not set, ATCP presumes that the loss is due to bit errors and simply retransmits the lost packet by entering the *retransmission state.*

TCP-BuS. This modification also uses network layer feedbacks to discover route failures [12]. This protocol exclusively chooses the associativity-based routing (ABR) protocol [13]. Two communication messages are used for maintenance of the TCP connection: explicit route disconnection notification (ERDN) and explicit route successful notification (ERSN). These messages are introduced to notify the TCP sender of route failure and route reestablishment, respectively. The node that identifies path breakdown is known as a pivoting node (PN). It sends an ERDN message to the TCP sender upon discovering a link failure. PNs use localized queries (LQs) to reestablish the route. After route reestablishment, a PN forwards an ERSN message to the source. After getting an ERSN, the TCP source continues its data broadcasting. This protocol is better than TCP-F and TCP-ELFN, but its execution depends on the underlying routing protocol, and it requests a buffering potential at intermediate nodes.

1.3.3.3 Non-TCP

ACTP. The application controlled transport protocol (ACTP) is lightweight and not an extension of TCP [14]. Unlike UDP, it provides feedback to the application regarding the status of the connection(s). ACTP supports the priority of packets to be sent, but it is the responsibility of lower layers to actually provide a differentiated service based on priority. It is executed as a layer between the application layer and the network layer. The application layer uses APIs to connect with the ACTP layer. It is scalable for larger networks. The protocol allows applications complete control in deciding the level of reliability and the quality of service (QoS) desired for different portions of a data stream.

Throughput is not affected by path breakdowns, but it is not compatible with TCP. When it is used in large ad-hoc networks, it can detect heavy congestion, but does not provide any congestion control mechanisms.

ATP. The ad-hoc transport protocol (ATP) [15] is designed to overcome the limitation shown by TCP. It is the antithesis of TCP. It differs from TCP as it coordinates among multiple layers and has a rate-based transmission. Congestion in TCP causes TCP connections to enter the slow-start mode frequently and several times. ATP uses a quick-start mechanism to deal with congestion. A three-phase rate adaptation technique minimizes packet losses due to congestion. ATP has a coarse-grained receiver feedback unlike TCP, which depends on ACKs.

1.3.3.4 Comparison of Various Protocols End-to-end protocols necessitate adjustments or modifications to the existing TCP codes at the mobile hosts or the fixed hosts. Compatibility between the network nodes is violated. This in turn requires recompilation and linking of applications presently executing on the fixed hosts. This is a major drawback of any end-to-end protocol. Split connection protocols, on the other hand, have backward compatibility with the offered wired network protocol. No adjustments are required at the fixed hosts for accommodating mobile hosts.

Non-TCPs like ATP provide better performance than TCP. They help in decoupling of congestion control and reliability mechanisms. Non-TCPs also show improvement in avoidance of congestion window fluctuations. They exhibit better performance than default TCP, TCP-ELFN, and ATCP.

1.4 Security Issues: Threats and Mitigation Potentials

Wireless ad-hoc networks are formed by network nodes that have the ability to organize themselves into a dynamic, arbitrary network topology in the absence of any standard infrastructure. Furthermore, the wireless nodes may have mobility, thus allowing people and vehicles with network devices to connect even without any existing infrastructure. Wireless nodes can listen and communicate with all other wireless nodes in their radio range. Distant nodes communicate via intermediate hops. Wireless ad-hoc networks with mobile nodes have the following features [1]:

- *Unreliable wireless medium.* Wireless communication channels are more exposed to the environment than guided media. This exposes them to noise and other vulnerabilities. Also, mobility of the wireless nodes causes the wireless links to become inconsistent for communication purposes.
- *Dynamic topologies.* As the mobile, wireless nodes travel into and out of the radio range of other nodes within the network, the topology undergoes a constant change. This in turn results in routing information also being changed at each node.
- *Security loopholes.* Wireless ad-hoc routing protocols are not adaptable to the dynamically changing environment of wireless ad-hoc networks if designed statically. This requires the wireless nodes to incorporate security add-ons that plug into the underlying routing protocol. Adjacent nodes need to incorporate these changes for safeguarding themselves from potential vulnerabilities and attacks that may result from statically configured routing protocols.

The issues increase the vulnerabilities of wireless ad-hoc network environments. The wireless nodes are prone to a larger number of security threats than their wired counterparts. This demands extensive research in the domain of securing wireless ad-hoc network environments.

1.4.1 *Vulnerabilities of the Mobile Ad-Hoc Networks*

Securing the wireless ad-hoc networking environment is much more difficult than securing wired networks, as ad-hoc networks with wireless mobile nodes have greater security vulnerabilities. The following vulnerabilities should always be kept in mind while proposing security solutions for this environment.

1.4.1.1 Absence of Secure Boundaries Compared to wired networks, attackers in a wireless ad-hoc environment do not need to gain physical access to a link for joining the network. The very ad-hoc nature of the network allows nodes (or adversaries) to become a part of the network whenever they come within the radio range of any participating node [2]. Unauthorized physical access to wired networks requires hacking into several lines of defense, such as firewalls and gateways. The absence of any such secure boundary makes the wireless ad-hoc network environment more prone to attacks.

Direct access to the wireless link makes a wireless ad-hoc network susceptible to various types of attacks, such as passive eavesdropping, active interfering, leakage of secret information, data tampering, message replay, message contamination, and denial of service.

1.4.1.2 Malicious Insiders A different domain of attacks arises when wireless nodes are compromised to behave maliciously rather than modifying the wireless link. The term malicious insiders refers to those nodes that have been compromised for behaving malignantly. Malicious behavior can be classified using either *signature-based detection* or *anomaly-based detection.* In either situation, it becomes very difficult to distinguish abnormal behavior from malignancy, as the behavior of different nodes of the wireless ad-hoc network may be diverse. Mobility aids the attacker in that it can change its point of intrusion into the network quite frequently. Detecting such malicious nodes becomes all the more difficult in large-scale networks, especially because such nodes behave in a benign manner before being compromised. A Byzantine failure is a classical example of such an attack, where subsets of wireless nodes launch a synergistic attack and remain undetected by other nodes. Cooperation among malicious insiders can prove to be quite harmful for a wireless ad-hoc networking environment.

1.4.1.3 No Centralized Management Facility Network administration becomes all the more difficult for the wireless ad-hoc environment, as the very architecture of wireless networks is distributed. This results in added vulnerabilities, as follows:

- Monitoring the traffic, the wireless nodes, and the wireless link becomes very difficult in a large-scale wireless ad-hoc network. Time-varying behavioral patterns need to be analyzed for identifying misbehavior. This can be easily done by a central server that monitors the entire networking environment. Behavioral patterns over short periods of time, as observed by the wireless nodes of a dynamically changing network topology, are not sufficient to distinguish benign failures from malicious ones [3].

- Establishing a secure line of defense is not possible without a centralized architecture. An initial classification of untrusted and trustworthy nodes is not possible for establishing a security boundary [1].
- Cooperation among the wireless nodes becomes mandatory for executing certain specific protocols that are specifically designed for the mobile wireless ad-hoc networking environment. This decentralized requirement may be exploited by intruders to launch collaborative attacks that affect network performance [2].

1.4.1.4 Energy Constraints Nodes in a wired network run on electricity coming out from power outlets. So there is no energy constraint as such. The situation completely changes for wireless nodes that are on the move. These nodes run on battery and depend on the battery life for proper functioning. When the battery charge empties, it becomes mandatory to recharge the batteries. Replenishing the battery charge is not always possible when a wireless node is on the move. This causes several problems.

Two primary issues associated with the constrained energy of wireless nodes are denial of service (DoS) and selfish behavior. A DoS attack can be easily launched on a wireless node by overburdening it with useless work such as routing of infinitely many dummy packets, etc. Such an attack depletes the charge stored in the batteries of wireless nodes, and thus denies other wireless nodes of the network from services offered by the target node. Selfish behavior, on the other hand, is not always malicious. A wireless node tries to conserve its energy resources, especially when they fall below a critical threshold. Under such a situation the node may avoid cooperating with other nodes of the network for proper functioning of different wireless ad-hoc algorithms. If such noncooperation is intentional, then the selfishness is malicious.

1.4.1.5 Scalability One of the most attractive features of wireless ad-hoc networks is scalability. Since wireless nodes form the network on an ad-hoc basis, different nodes can join the network on the fly. This has large implications, as the size of the network can range from tens to hundreds to thousands of nodes. All network management algorithms, such as routing, access control, and key distribution, must also be adaptable to such changes. This particular feature of wireless ad-hoc networks has a severe impact on the security threats that the network gets exposed to.

1.4.1.6 Summary It is evident that wireless mobile ad-hoc networks are insecure by nature.

1. There is no clear line of defense because of the freedom for the nodes to join, leave, and move inside the network.
2. Some of the nodes may be compromised by the adversary, and thus perform some malicious behaviors that are hard to detect.
3. Lack of centralized machinery may cause some problems when there is a need to have such a centralized coordinator.
4. Restricted power supply can cause some selfish problems.
5. Continuously changing the scale of the network has set a higher requirement to the scalability of the protocols and services in the mobile ad-hoc network.

Thus, compared with wired networks, the wireless mobile ad-hoc network will need more robust security schemes. Some of these are discussed in the next section.

1.4.2 Mitigation Potentials of Wireless Ad-Hoc Networks

A detailed discussion has already been presented on the vulnerabilities of wireless ad-hoc networks made up of wireless mobile nodes. However, it is not sufficient that we merely recognize the imminent threats that impinge upon the normal operations of a wireless ad-hoc network. Research needs to be done to develop security schemes for wireless mobile ad-hoc networks. The following sections systematically explore this domain.

1.4.2.1 Security Criteria It is essential to define a secure configuration for a wireless ad-hoc network. In other words, what are the security criteria that, if fulfilled, will ensure that the resulting wireless ad-hoc network is secure? Such a question can be answered only by defining certain parameters for a secure wireless ad-hoc network. The security criteria are as follows:

Availability. It is the ability of a node to deliver the entire set of network-specific services irrespective of its security stature. DoS attacks and selfish behavior may disrupt network services [1, 2].

Integrity. Integrity ensures that messages have not been tampered with during transmission. Messages can be dropped, replayed, or modified intentionally with malicious purposes. They may also undergo change due to some nonmalignant failure such as a hardware crash [4].

Confidentiality. Messages should be accessed only by those who have the authorization to access them. Access privileges may prevent other nodes from reading the contents of confidential messages.

Authenticity. It is a mechanism to sever impersonation attacks during communication. Both sender and receiver of messages must prove their identities using some authentication mechanism during message transfer [1].

Nonrepudiation. It ensures that the sender and receiver cannot disown the messages they have sent or received. Thus, compromised nodes sending erroneous messages can be identified if nonrepudiation is ensured.

Authorization. These techniques provide privileges and permissions to the nodes within a network. Authorization is used to issue different access rights to different levels of users.

Anonymity. Privacy of the nodes is ensured by anonymity. The identity of the owner of a node is not distributed by the node or the system software.

1.4.2.2 Types of Intrusion in Wireless Ad-Hoc Networks Intrusions in wireless ad-hoc networks can be broadly classified into two different categories [2]:

1. *External attacks* where intruders disrupt the normal services of a network by causing congestion, falsified routing, or blocking other resources of the network.

2. *Internal attacks* imply that intruders gain normal access to the network by impersonating a good node or compromising it, and then launching specific attacks on the network.

Internal attacks are more severe than external attacks in that the former allows intruders to use the network resources with the same access privileges as a normal well-behaved node. These attacks are of greater concern when designing security solutions for wireless ad-hoc networks. The main types of attacks that occur in wireless ad-hoc networks are as follows:

DoS. Distributed jamming of the wireless link and battery exhaustion techniques are used by DoS agents to block network resources and thereby prevent the network from delivering the desired services.

Impersonation. Lack of proper authentication mechanisms can allow intruders to join a wireless ad-hoc network with the same access rights as a normal node. Network services may start malfunctioning as the intruders launch specific attacks on the network [1].

Eavesdropping. This type of attack tries to decrypt confidential information from a wireless communication. This information may include passwords and public or private keys.

Attacks against routing. Intruders may launch attacks on the routing protocol itself or on the packet-forwarding mechanism. The first type of attack tends to tamper with the routing information that is maintained by the wireless nodes. They include network partitioning, routing loop, resource deprivation, and route hijack. More complex attacks on the routing protocol include wormhole attacks, rushing attacks, and Sybil attacks. The second type of attack tries to misguide packets along other paths while being delivered to the destination. Selfish behavior and denial of service are two mechanisms of launching these types of attacks.

1.4.2.3 Security Approaches in Wireless Ad-Hoc Networks The vulnerabilities of wireless ad-hoc networks allow intruders to launch different types of attacks on them. Ensuring the security criteria of wireless mobile ad-hoc networks is essential for proper deliverance of network services. The security schemes that are popularly used to handle different types of attacks are as follows:

Intrusion detection techniques. An intrusion detection system (IDS) does not ensure that intrusions will never occur in the network. Rather, they secure a network by detecting unwanted manipulations that render network services obscured. Intrusions are malicious behaviors that need to be identified and separated from both normal and abnormal behavior. Abnormal behavior may occur under specific network configurations [5, 6]. However, it is not necessary that abnormal behavior is always malicious. It is very difficult to draw the line between the two. This is what gives rise to *false positives* and *false negatives* as security metrics for evaluating the performance and efficiency of an IDS.

- *Intrusion detection schemes for wireless ad-hoc networks.* Intrusions, in general, are defined as deviations from standardized normal behavior. The process of intrusion detection involves four different tasks.
  1. *Temporal data collection.* This is the most basic task that is performed by all nodes in the network. Each node monitors the behavior of its one-hop neighbors and maintains statistics for them.
  2. *Local detection engine.* Every node in the network analyzes the statistical data collected over a time period for intrusions. Deviations from normal behavior can be detected using *anomaly-based detection* or *signature-based detection.* Both these techniques try to detect malicious behavior.
  3. *Cooperative detection engine.* Sometimes it becomes difficult for a node to classify abnormal behavior strictly as malicious. Such suspicious behavioral data are shared among other wireless nodes of the network, and a decision is reached by consensus.
  4. *Intrusion response.* This phase defines how the network reacts after an intrusion is detected. Traffic routes may be regenerated, routing tables may be updated, suspected nodes may be blacklisted, or the public and private keys of the nodes may have to be regenerated and redistributed for the entire network.

- *Cluster-based intrusion detection technique for wireless ad-hoc networks.* A cluster is defined by a cluster head and all nodes that are within its radio range. A cluster-based IDS requires that every node within the network must be a part of a cluster. A cluster head is not a special node and has the same properties and access privileges as any other node in the network. However, the cluster head election mechanism must be *fair* and *efficient.* Fairness implies that all nodes within a cluster have the same probability of being selected as the cluster head, and that all members of the cluster must remain in the cluster for the same amount of time. Efficiency demands the cluster head selection procedures to execute periodically and with high efficiency [16].
- *Cross-layer intrusion detection techniques for wireless ad-hoc networks.* There is scope for ample research in the domain of cross-layer solutions for intrusion detection. Every day attackers are coming up with novel attack launching mechanisms. Multilayer attacks exploit the vulnerabilities of different layers and launch the attack partially from each of these layers. As a result, single-layer intrusion detection techniques fail to detect these types of advanced attacks. Cross-layer solutions combine behavioral data from different layers and analyze these data in a comprehensive manner [6, 17].

Secure routing techniques in mobile ad-hoc networks. As discussed previously, there are numerous kinds of attacks against the routing layer in the mobile ad-hoc networks, some of which are more sophisticated and harder to detect than others. Multiple solutions exist for securing the routing mechanism of mobile ad-hoc networks. Most of these solutions secure the network from specific attacks, such as *selective forwarding,* or *wormhole attacks. Watchdog* and *Pathrater* are two main components of a system that aim to mitigate the routing misbehaviors in mobile ad-hoc networks. *Self-healing* is a more recent concept whereby networks not only detect intrusions, but also respond to them by rerouting packets through intrusion-free paths or zones [18–21].

1.4.2.4 Summarizing the Mitigation Potentials of Wireless Ad-Hoc Networks
Intrusion detection is a reactive mechanism where the cost and overhead of securing the network are less. In this mechanism, intrusions are allowed to occur within the system and then are systematically detected and removed. On the other hand, securing the routing mechanism is a proactive mechanism where intruders or attackers are prevented from entering the network. This is more difficult to implement, as novel attacking schemes are evolving every day. However, both these approaches have their implementation scenarios and can be successfully combined to produce secure wireless mobile ad-hoc networking environments. There is ample scope of research for both approaches.

References

1. Amitabh Mishra and Ketan M. Nadkarni, Security in Wireless Ad hoc Networks, in *The Handbook of Ad hoc Wireless Networks*, chap. 30, CRC Press, Boca Raton, FL, 2003.
2. Yongguang Zhang and Wenke Lee, Security in Mobile Ad-Hoc Networks, in *Ad hoc Networks Technologies and Protocols*, chap. 9, Springer, Berlin, 2005.
3. Panagiotis Papadimitraos and Zygmunt J. Hass, Securing Mobile Ad hoc Networks, in *The Handbook of Ad hoc Wireless Networks*, chap. 31, CRC Press, Boca Raton, FL, 2003.
4. Wikipedia, Data Integrity, http://en.wikipedia.org/wiki/Data_integrity.
5. Wikipedia, Intrusion-Detection System, http://en.wikipedia.org/wiki/Intrusion-detection_system.
6. Y. Zhang and W. Lee, Intrusion Detection in Wireless Ad-Hoc Networks, in *Proceedings of the 6th International Conference on Mobile Computing and Networking (MobiCom 2000)*, Boston, August 2000, pp. 275–283.
7. A. Bakrey and B.R. Badrinath, Indirect TCP for Mobile Host, in *Proceedings of the 15th International Conference on Distributed Computing System*, May 30–June 2, 1995, p. 136.
8. K. Brown and S. Singh, M-TCP: Mobile TCP for Mobile Cellular Networks, in *Proceedings of the ACM SIGCOMM*, October 1997, vol. 27, pp. 19–43.
9. K. Chandran, S. Raghunathan, S. Venkatesan, and R. Prakash, A Feedback Based Scheme for Improving TCP Performance in Ad hoc Wireless Networks, in *Proceedings of the International Conference on Distributed Computing Systems*, May 1998, pp. 472–479.
10. G. Holland and N.H. Vaidya, Analysis of TCP Performance over Mobile Ad hoc Networks, in *Proceedings of the ACM MOBICOM Conference*, August 1999, pp. 219–230.

11. J. Liu and S. Singh, ATCP: TCP for Mobile Ad hoc Networks, *IEEE Journal on Selected Areas in Communications*, 19(7), 1300–1315, 2001.
12. D. Kim, C. Toh, and Y. Choi, TCP-Bus: Improving TCP Performance in Wireless Ad hoc Networks, *Journal of Communications and Networks*, 3(2), 175–186, 2001.
13. C. Toh, Associativity-Based Routing for Ad-Hoc Mobile Networks, *Journal of Wireless Personal Communications*, 4(2), 130–139, 1997.
14. J. Liu and S. Singh, ACTP: Application Controlled Transport Protocol for Mobile Ad-Hoc Networks, in *Proceedings of IEEE WCMC 1999*, September 1999, vol. 3, pp. 1318–1322.
15. K. Sundaresan, V. Anantharaman, H.-Y. Hsieh, and R. Sivakumar, ATP: Reliable Transport Protocol for Ad-Hoc Networks, in *Proceedings of 4th ACM International Symposium on Mobihoc*, May 2003.
16. Yi-an Huang and Wenke Lee, A Cooperative Intrusion Detection System for Ad hoc Networks, in *Proceedings of the 1st ACM Workshop on Security of Ad hoc and Sensor Networks*, Fairfax, VA, 2003, pp. 135–147.
17. Jim Parker, Anand Patwardhan, and Anupam Joshi, Detecting Wireless Misbehavior through Cross-Layer Analysis, in *Proceedings of the IEEE Consumer Communications and Networking Conference Special Sessions (CCNC 2006)*, Las Vegas, NV, 2006.
18. Y. Hu, A. Perrig, and D. Johnson, Packet Leashes: A Defense against Wormhole Attacks in Wireless Ad hoc Networks, in *Proceedings of IEEE INFOCOM'03*, 2003.
19. Sergio Marti, T.J. Giuli, Kevin Lai, and Mary Baker, Mitigating Routing Misbehavior in Mobile Ad hoc Networks, in *Proceedings of the 6th Annual International Conference on Mobile Computing and Networking (MobiCom'00)*, Boston, 2000, pp. 255–265.
20. Jiejun Kong, Xiaoyan Hong, Yunjung Yi, JoonSang Park, Jun Liu, and Mario Gerlay, A Secure Ad-Hoc Routing Approach Using Localized Self-Healing Communities, in *Proceedings of the 6th ACM International Symposium on Mobile Ad hoc Networking and Computing*, Urbana-Champaign, IL, 2005, pp. 254–265.
21. Y. Hu, A. Perrig, and D. Johnson, Wormhole Attacks in Wireless Networks, *IEEE Journal on Selected Areas in Communications*, 24(2), 2006.
22. IEEE Standard Association, *IEEE Standard for Telecommunications and Information Exchange between Systems—LAN/MAN—Specific Requirements—Part 15: Wireless Medium Access Control (MAC) and Physical Layer (PHY) Specifications for Wireless Personal Area Networks (WPANs)*, http://standards.ieee.org/findstds/standard/802.15.1-2002.html.
23. IEEE 802.15 WPAN Task Group 4 (TG4), *IEEE Std. 802.15.4-2003, IEEE Standard for Information Technology—Telecommunications and Information Exchange between Systems—Local and Metropolitan Area Networks—Specific Requirements—Part 15.4: Wireless Medium Access Control (MAC) and Physical Layer (PHY) Specifications for Low Rate Wireless Personal Area Networks (WPANs)*, IEEE Press, New York, 2003, http://www.ieee802.org/15/pub/TG4.html.

24. Jose A. Gutierrez, Edgar H. Callaway Jr., and Raymond L. Barrett Jr., *Low-Rate Wireless Personal Area Networks: Enabling Wireless Sensors with IEEE 802.15.4™*, IEEE Press, New York, 2003.
25. Ed Callaway, *Low Power Consumption Features of the IEEE 802.15.4/ ZigBee LR-WPAN Standard*, Technical Report, Motorola Labs, 2003, http://www.cens.ucla.edu/sensys03/sensys03-callaway.pdf.
26. Shahin Farahani, *ZigBee Wireless Networks and Transceivers*, Newnes, Newton, MA, 2008.
27. Brent A. Miller, Chatschik Bisdikian, and Tom Siep, *Bluetooth Revealed*, Prentice Hall, Upper Saddle River, NJ, 2001.
28. Sanjeev Sharma, R.C. Jain, Sarita Bhadauria, and Ramesh Cherukuru, Comparative Study of MAC Protocols for Mobile Ad Hoc Networks, *International Journal of Soft Computing*, 1, 2006, pp. 225–231.

2

Architecture and Organization Issues

MANALI CHAKRABORTY, NOVARUN DEB, DEBDUTTA BARMAN ROY, AND RITUPARNA CHAKI

Contents

2.1 Introduction

The first communication without any human intervention was introduced by George Stibitz in 1940 using a teletype to send instructions from Dartmouth College to New York. Before that, communications among machines were rarely performed, using hand-carried removable storage media. After 1940, a gradual development was seen in the history of the computer network [1].

A network is conceived of as a group of computers logically connected for the sharing of information or services [2]. The sharing of information is done in various ways, like print services, multitasking, sharing of files, etc. Though initially file sharing and printer sharing were the main purposes of networks, later on they have been used in application sharing and business logic sharing. Networks are broadly classified into wired networks and wireless networks. A taxonomy showing the different kinds of wireless networks is presented in Figure 2.1.

This book primarily focuses on infrastructure-less wireless ad-hoc networks—how they have evolved over time and how their characteristic features have been exploited by attackers for launching different types of attacks. Solutions for dealing with these attacks are also presented in this book. This chapter also has a discussion on hybrid

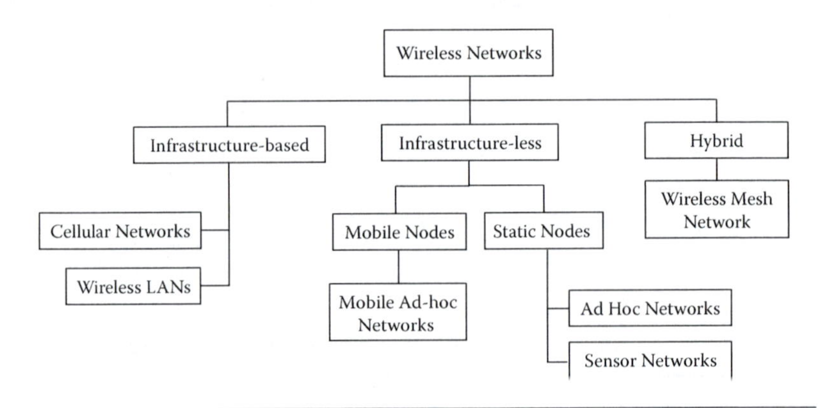

Figure 2.1 Taxonomy of wireless networks.

wireless networks, namely, wireless mesh networks, which is the current trend in network architecture and organization.

2.2 Wireless Ad-Hoc Networks (WANs)

After an era of providing solutions in the domain of infrastructure-based wired networks, several commercial applications cropped up that required providing services to clients on the go. This basic need led to the development of wireless ad-hoc networks (Figure 2.2). Once protocols and standards were developed for WANs, the need for security became obvious.

To protect networks from adversaries, security issues in ad-hoc networks (AHNs) are investigated based on the knowledge of security measures in wired networks. AHNs were prone to the same types of attacks as wired networks. Furthermore, the openness of wireless communication media made AHNs more vulnerable to attacks than traditional networks. Anyone with a scanner could monitor traffic from the comfort of his or her home or the ease of a street corner. With a powerful jamming machine, an attacker could reduce the channel availability or even shut down communication channels [4].

Wired networks were built over time. They reflected security policies of organizations. Trust between entities, an essential element of security policy, was also built over time. System administrators supported network operations such as implementing security policies. In comparison,

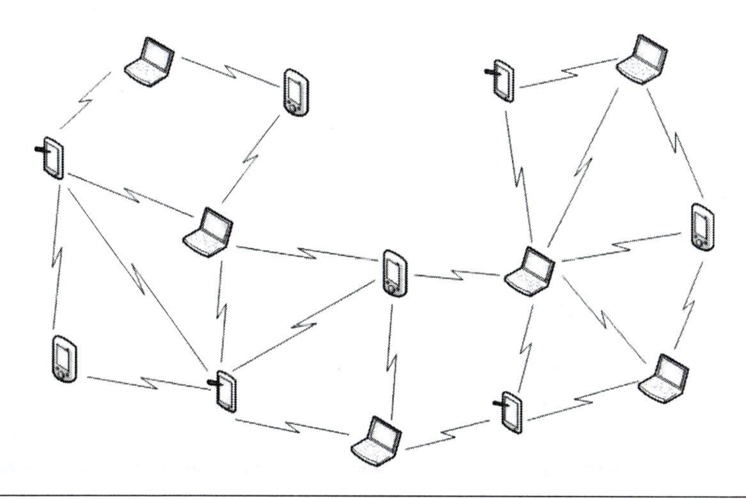

Figure 2.2 Wireless ad-hoc network.

AHNs were built quickly and as needed. Trust and policies were put together in a hurry. Some physical features (e.g., small size) of nodes made them more easily compromised and lost than those in wired networks.

Different AHNs have different initial contexts and requirements for security, depending on applications. However, they all share one characteristic: no fixed infrastructure. The lack of infrastructure support led to the absence of dedicated machines providing naming and routing services. Every node in an AHN became a router. Thus, network operations had higher dependence on individual nodes than on wired networks. The ad-hoc nature of the nodes brought constant change in network topology and membership, making it impractical to provide traditional, centralized services [4, 5].

The unreliability of wireless links between nodes, the constantly changing topology owing to the movement of nodes in and out of the network, and the lack of incorporation of security features in statically configured wireless routing protocols not meant for ad-hoc environments all led to increased vulnerability and exposure to attacks. Security in wireless ad-hoc networks was particularly difficult to achieve, notably because of the vulnerability of the links, the limited physical protection of each of the nodes, the sporadic nature of connectivity, the dynamically changing topology, the absence of a certification authority, and the lack of a centralized monitoring or management point [6]. This, in effect, underscored the need for intrusion detection, prevention, and related countermeasures.

The absence of infrastructure and the consequent absence of authorization facilities impeded the usual practice of establishing a line of defense, distinguishing nodes as trusted and nontrusted. Such a distinction would have been based on a security policy, the possession of the necessary credentials, and the ability of nodes to validate them. In the case of wireless ad-hoc networks, there might have been no grounds for an a priori classification, since all nodes were required to cooperate in supporting the network operation, while no prior security association could be assumed for all the network nodes [7].

2.3 Mobility of Nodes and MANETs

Mobile ad-hoc networks (MANETs) were vulnerable to a wide range of active and passive attacks that could be launched relatively

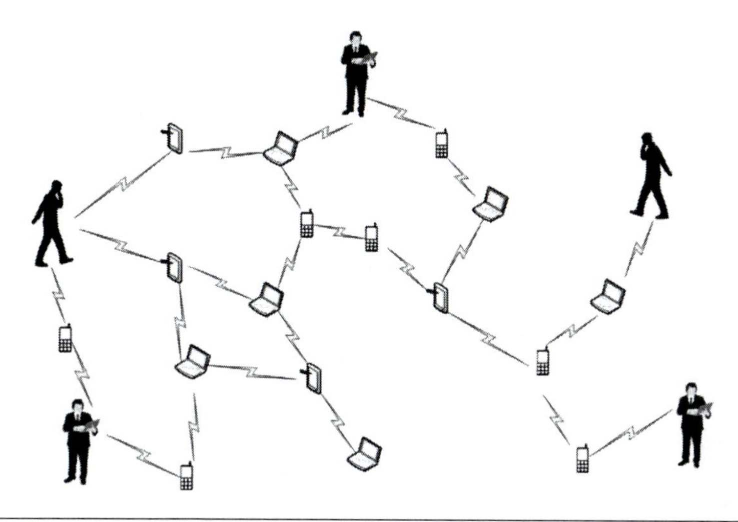

Figure 2.3 Mobile ad-hoc network.

easily, since all communications take place over the wireless medium (Figure 2.3). In particular, wireless communication facilitates eavesdropping, especially because continuous monitoring of the shared medium, referred to as promiscuous mode, was required by many MANET protocols. Impersonation was another attack that became more feasible in the wireless environment. Physical access to the network was gained simply by transmitting with adequate power to reach one or more nodes in proximity, which may have no means to distinguish the transmission of an adversary from that of a legitimate source. Finally, wireless transmissions could be intercepted, and an adversary with sufficient transmission power and knowledge of the physical and medium access control layer mechanisms could obstruct its neighbors from gaining access to the wireless medium.

In addition, freely roaming nodes join and leave MANET subdomains independently, possibly frequently and without notice, making it difficult in most cases to have a clear picture of the ad-hoc network membership. In other words, there may be no grounds for an a priori classification of a subset of nodes as trusted to support the network functionality. Trust may only be developed over time, while trust relationships among nodes may also change, when, for example, nodes in an ad-hoc network dynamically become affiliated with administrative domains. This was in contrast to other mobile networking paradigms, such as mobile Internet Protocol (IP) or cellular telephony, where nodes continue to belong to their administrative domain in spite of mobility. Consequently, security solutions with static configuration would not suffice, and the assumption that all nodes can be bootstrapped with the credentials of all other nodes would be unrealistic for a wide range of MANET instances [7, 8].

The absence of a central entity made the detection of attacks a very difficult problem, since highly dynamic large networks cannot be easily monitored. Benign failures, such as transmission impairments, path breakages, and dropped packets, were naturally a fairly common occurrence in mobile ad-hoc networks, and consequently, malicious failures would be more difficult to distinguish. This will be especially true for adversaries that vary their attack pattern and misbehave intermittently against a set of their peers that also changes over time. As a result, short-lived observations would not allow detection of adversaries.

2.3.1 *Mobility of Nodes in Mobile Ad-Hoc Network*

Mobility is the basic parameter for every node in a mobile ad-hoc network. To describe the movement pattern of mobile nodes, the mobility model is designed. This model also describes the changes in velocity, location, and acceleration over time. Due to the importance of mobility, it is desirable for mobility models to emulate the movement pattern of targeted real-life applications in a reasonable way [11, 12] (Figure 2.4).

2.3.1.1 Random-Based Mobility Models In random-based mobility models, the mobile nodes are free to move in any direction and any destination at random speed, independent of other nodes in the network. This kind of model has been used in many simulation studies.

2.3.1.2 Mobility Models with Temporal Dependency Mobility of a node may be constrained and limited by the physical laws of acceleration, velocity, and rate of change of direction. Hence, the current velocity of a mobile node may depend on its previous velocity. Thus, the velocities of a single node at different time slots are correlated. We call this mobility characteristic the temporal dependency of velocity

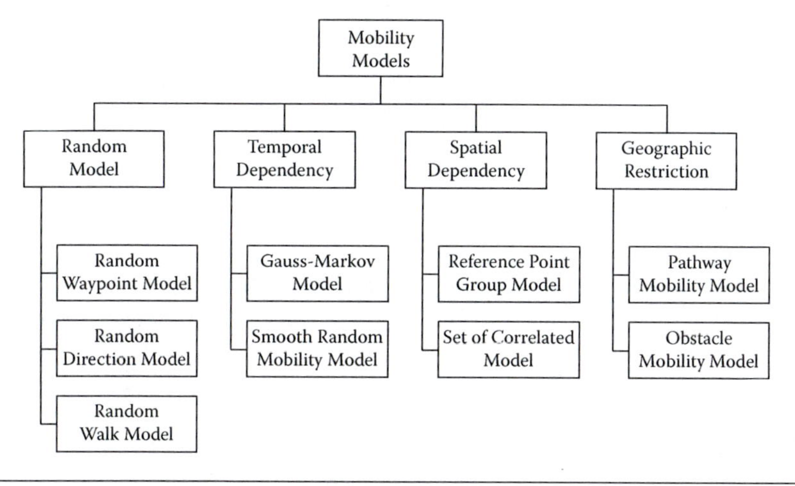

Figure 2.4 Taxonomy of mobility model.

2.3.1.3 Mobility Models with Spatial Dependency In the random waypoint model and other random models, a mobile node moves independently of other nodes in any direction with any speed. In some targeted MANET applications, team collaboration among users exists, and the users are likely to follow the team leader. Therefore, the mobility of the mobile node could be influenced by other neighboring nodes. Since the velocities of different nodes are correlated in space, this feature is called spatial dependency of velocity.

2.3.1.4 Mobility Models with Geographic Restriction Mobile nodes, in the random waypoint model, are allowed to move freely and randomly anywhere in the simulation field. However, in most real-life applications, it is observed that the mobility of a node depends on the environment and obstacles that exist in the environment. This kind of mobility model is called a mobility model with geographic restriction.

Abnormal situations occurred frequently because nodes behaved in a selfish manner and did not always assist the network functionality. It was noteworthy that such behavior may not be malicious, but only necessary when, for example, a node shuts its transceiver down in order to preserve its battery [8]. Thus, from obvious reasoning, it can be anticipated that providing an infrastructure to ad-hoc networks has become the need of the hour. The next advancement in networking environments was sensor networks.

2.4 Sensor Networks

Wireless sensor networks are quickly gaining popularity due to the fact that they are potentially low-cost solutions to a variety of real-world challenges [9] (Figure 2.5). Their low cost provides a means to deploy large sensor arrays in a variety of conditions capable of performing both military and civilian tasks. But sensor networks also introduce severe resource constraints due to their lack of data storage and power. Both of these represent major obstacles to the implementation of traditional computer security techniques in a wireless sensor network. The unreliable communication channel and unattended operation make the security defenses even harder. Indeed, as pointed out in [10], wireless sensors often have the processing characteristics of machines that are decades old (or older), and the industrial trend is to reduce the cost of wireless sensors while maintaining similar computing power.

With that in mind, many researchers have begun to address the challenges of maximizing the processing capabilities and energy reserves of wireless sensor nodes while also securing them against

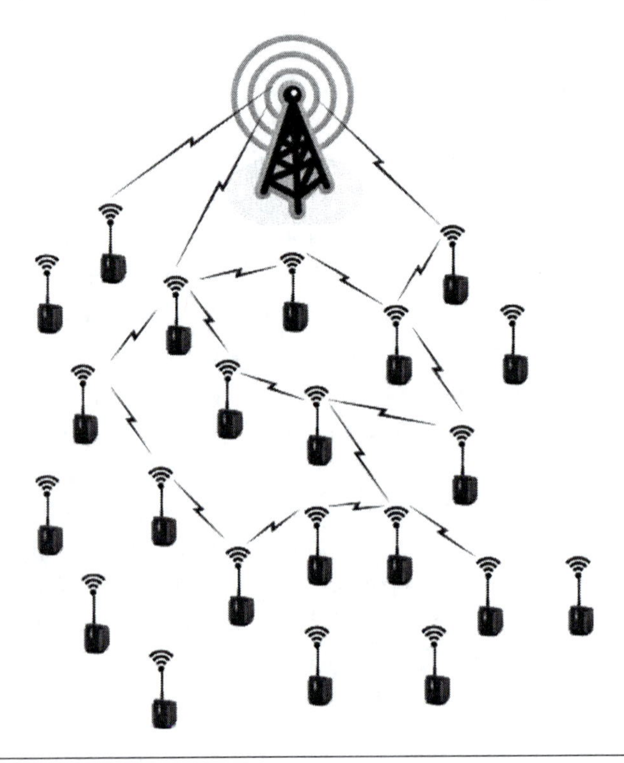

Figure 2.5 Sensor network.

attackers. All aspects of the wireless sensor network are being examined, including secure and efficient routing [8, 11–13], data aggregation [14–19], group formation [20–22], and so on.

In addition to those traditional security issues, we observe that many general purpose sensor network techniques (particularly the early research) assumed that all nodes are cooperative and trustworthy. This is not the case for most, or much of, real-world wireless sensor networking applications, which require a certain amount of trust in the application in order to maintain proper network functionality. Researchers therefore began focusing on building a sensor trust model to solve problems beyond the capability of cryptographic security [23–30]. In addition, there are many attacks designed to exploit the unreliable communication channels and unattended operation of wireless sensor networks. Furthermore, due to the inherent unattended feature of wireless sensor networks, we argue that physical attacks to sensors play an important role in the operation of wireless sensor networks.

2.5 Cluster and Hierarchy in Organization

The most essential characteristic of wireless ad-hoc networks is that they have no infrastructure. This implies that these types of networks are essentially peer-to-peer environments where all the nodes forming the network partially undertake the responsibility of performing the different network operations, like routing, etc. The basic and most critical assumption in this type of networking environment is that the participating nodes are somewhat *trustworthy*; otherwise, networking operations cannot be performed. Thus, collaboration between nodes or groups of nodes is mandatory for proper and efficient management of wireless ad-hoc networks.

Clustering is one such technique of grouping nodes that improves the energy efficiency and other performance metrics of such a network. There are two different models that define the architecture of clusters: the *flat* model and the *hierarchical* model. The flat model can be further classified into *physical* (or *real*) clustering and *logical* (or *virtual*) clustering. The hierarchical model can also be classified into two approaches: *rigid hierarchy* and *flexible hierarchy*. The classifications are presented in Figure 2.6.

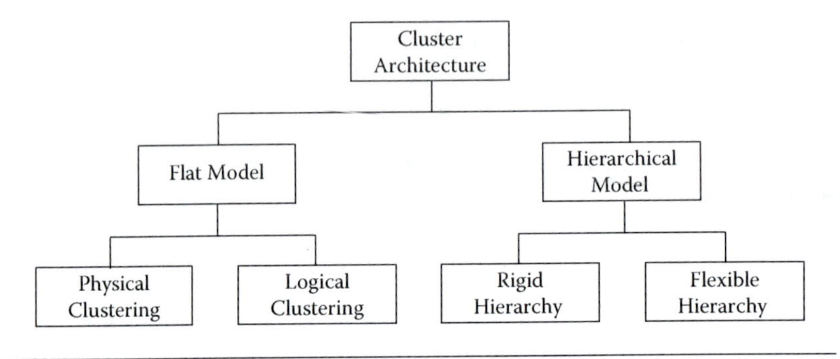

Figure 2.6 Classification of cluster architectures.

2.5.1 *The Flat Clustering Model*

In general, nodes in a wireless ad-hoc network are placed randomly and within radio range of each other. Being a peer-to-peer environment, all nodes in the network share the same responsibilities and perform the same set of functions that are required for the proper functioning of the network. However, the very concept of clustering results in the categorization of nodes into cluster members and cluster heads. Cluster heads are burdened with the additional duty of maintaining the clusters. Cluster members, on the other hand, behave as normal wireless nodes and communicate through their respective cluster heads.

In the *flat clustering model*, clusters are formed from all the wireless nodes within the network. All the network nodes are on the same level of communication; that is, all the nodes communicate over the same wireless medium using the same set of frequencies. However, once clusters are formed, the network may be perceived as a two-layer architecture with all the cluster heads communicating at the upper layer and all cluster members interacting from the lower layer.

2.5.1.1 Physical or Real Clustering Physical clustering is also referred to as near-term digital radio (NTDR) networks [3]. In this scheme, wireless nodes are clustered based on their physical location and radio range. A lot of research has been done on cluster head election techniques, and several algorithms exist that can efficiently elect cluster heads from groups of wireless nodes. Once cluster heads have been elected, it is their duty to broadcast their newly attained status to all one-hop neighbors. Non-cluster head nodes receive broadcast updates

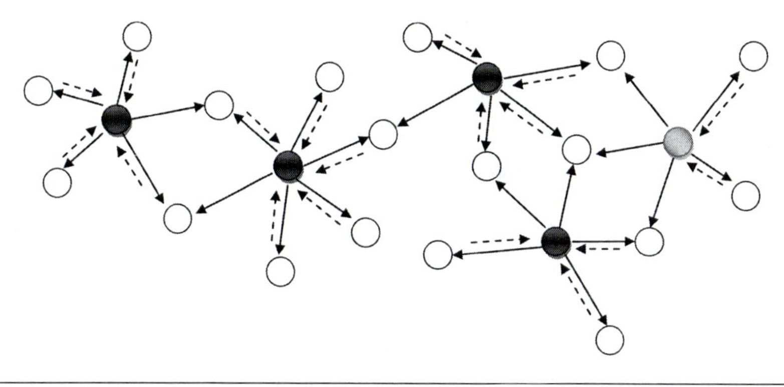

Figure 2.7 Physical or real clustering.

from one-hop neighbors who have been newly elected as cluster heads. On receiving such a HELLO packet, these nodes must join a cluster by confirming or returning an acknowledgment to the respective cluster head. If such a node receives multiple broadcasts from different cluster heads, it may either chose to become a member of any one cluster or act as a gateway between these clusters by joining them simultaneously. All nodes must behave as either cluster head or cluster member belonging to some cluster. If a node is not elected as the cluster head and it also does not receive any broadcasts from any neighboring cluster head, then it is not in the radio range of any of the elected cluster heads. In that case, such a node may declare itself a cluster head of a newly defined cluster. The process is clearly explained in Figure 2.7. The black cluster heads have been elected using some standard cluster head election algorithm. The gray cluster head is a self-proclaimed one that did not receive any broadcast message from any of the elected cluster heads.

The above-described model may apparently seem to be static, but it may be modified to incorporate dynamic network topologies that result from mobility of the nodes. Periodic execution of the cluster head election algorithm will result in readjusting the clusters after definite time slices.

Once clusters are formed, the routing of traffic becomes very intuitive. All intracluster communication takes place through the respective cluster heads. The channel frequency used by all the clusters within a network may be the same or different for different clusters. Further, the cluster heads at the upper layer form a backbone by communicating on a completely different band of the frequency spectrum. This is

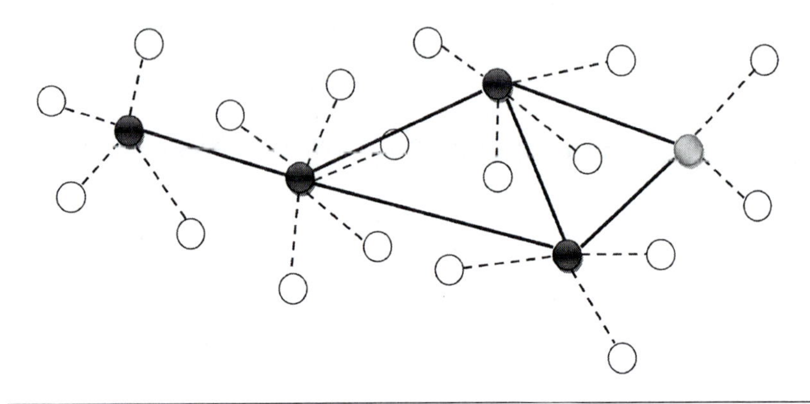

Figure 2.8 Physical clustering with backbone.

shown in Figure 2.8. The dotted lines represent intracluster communication links, and the bold lines represent the backbone communication network.

Intercluster communication is achieved when a cluster member sends a packet to its respective cluster head. The cluster head then forwards the packet through the backbone network to the destination cluster. Once the packet reaches the destination cluster, the cluster head of that cluster delivers the packet to the destination node. This mechanism can be easily interpreted from Figure 2.8.

2.5.1.2 Logical or Virtual Clustering This is a completely different clustering scheme for creating subnets from the wireless nodes forming a wireless ad-hoc network. Logical or virtual clustering is based on grouping nodes based on certain parameters or services rather than geographical location, as in physical clustering. Each virtual cluster is a collection of nodes that belong to the same service group or deal with the same service parameters.

Logical or virtual clustering eliminates the concept of cluster heads and introduces the concept of cluster gateways. Virtual clusters are formed over the wireless nodes of a network, but with the underlying assumption that the nodes have been geographically grouped into *physical subnets*. Physical subnets must not be confused with the physical clusters that were described in the previous section. Their main difference lies in the fact that physical subnets consist of one or more cluster gateways and cluster members, instead of a single cluster head presiding over a set of cluster members, as in physical clusters.

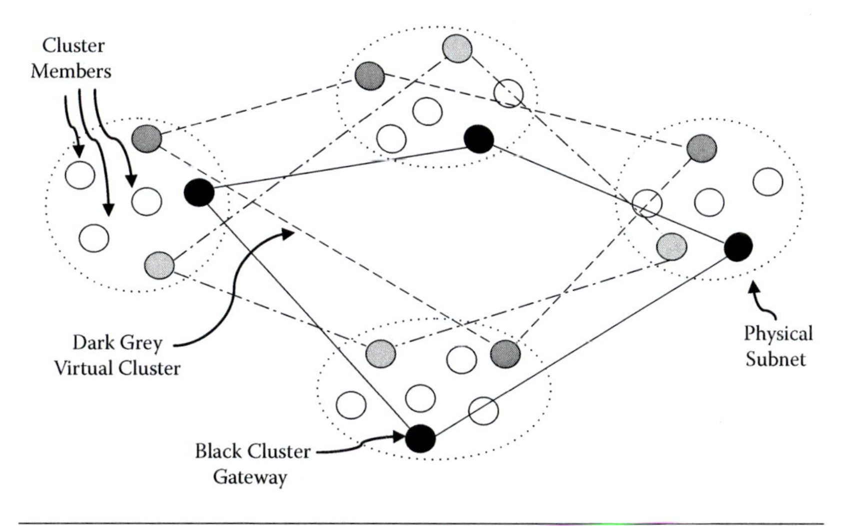

Figure 2.9 Virtual clustering over physical subnets.

A cluster gateway is very similar to a cluster member. The only difference is that a cluster gateway not only belongs to a physical subnet but also is a member of a virtual cluster. Thus, a virtual cluster is formed from participating cluster gateways belonging to different physical subnets. Each virtual cluster communicates over a different frequency band. A cluster gateway is so called because once a packet reaches a cluster gateway, it can hop from one physical subnet to another through other cluster gateways that belong to the same virtual cluster. Figure 2.9 illustrates the concept of logical clusters over physical subnets.

Routing packets using virtual clusters becomes very interesting. There are three different scenarios that need to be addressed:

1. *Implicit routing*: This is the situation when both the source and destination nodes are cluster gateways and belong to the same virtual cluster. Here the packets are forwarded along the virtual cluster from the source node to the destination node. The source and destination nodes may belong to two different physical subnets. Figure 2.10 illustrates this routing.
2. *Direct routing*: This routing scheme is required when either the source or the destination is a cluster gateway of some virtual cluster. The other node is an ordinary cluster member belonging to a physical subnet that contains a cluster gateway for

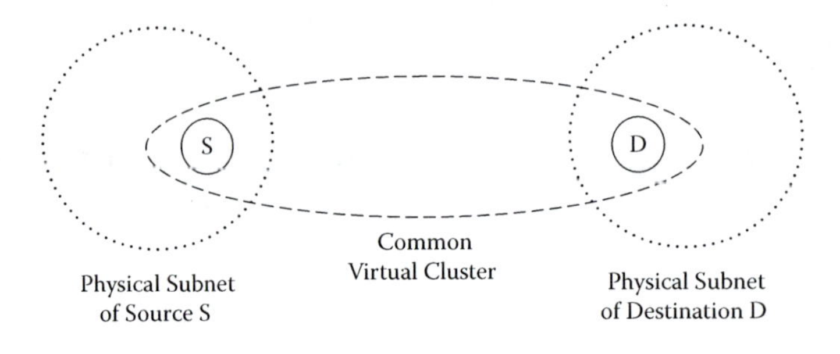

Figure 2.10 Implicit routing using virtual clusters.

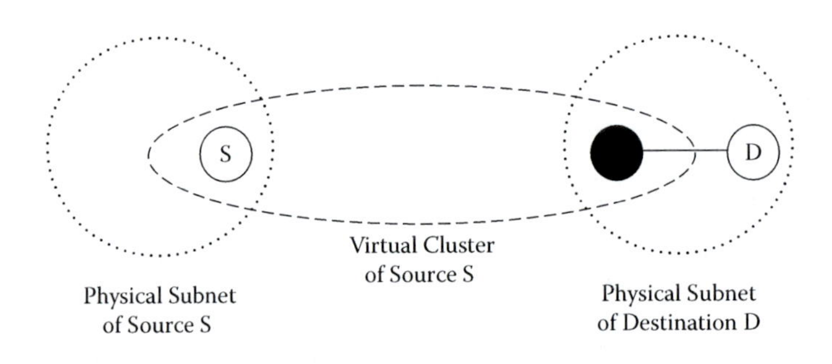

Figure 2.11 Direct routing using virtual clusters.

that virtual cluster. Let us consider the case when the source is a cluster member and the destination is a cluster gateway. Figure 2.11 illustrates the routing process.

Since the destination node is a cluster gateway of some virtual cluster, the source node reaches an intermediate cluster gateway within its physical subnet that belongs to the destination node's virtual cluster. Once that is done, the remaining process is similar to implicit routing. Figure 2.12 shows the other alternative of direct routing. The routing technique is intuitive and self-explanatory.

3. *Long path routing*: In this scenario, either both the source and destination nodes are ordinary cluster members, or they are both cluster gateways, but of different virtual clusters. In the latter case, long path routing will be done only when the source's virtual cluster has no cluster gateway in the destination's physical subnet and vice versa.

   The first situation is illustrated in Figure 2.13. Here both the source and destination nodes are ordinary cluster members. The source tries to find a virtual cluster that has gateways

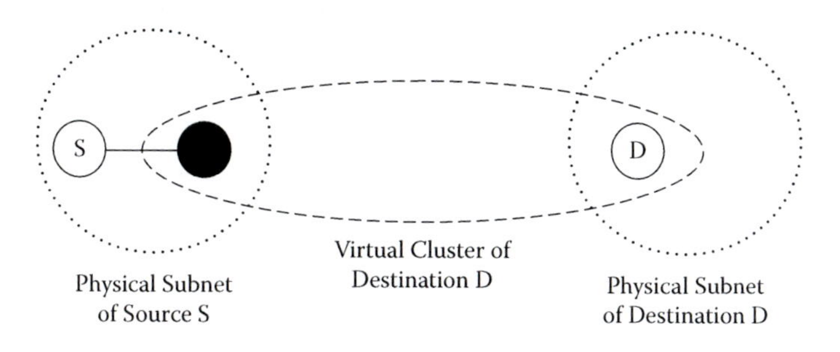

Figure 2.12 Direct routing using virtual clusters.

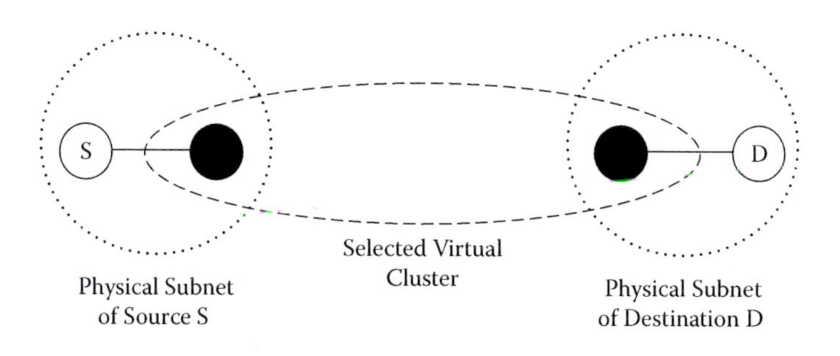

Figure 2.13 Long path routing using virtual clusters.

in both the source and the destination's physical subnets. Once that is established, the source forwards its packets to the desired intermediate cluster gateway. The packets then traverse the network through the desired virtual cluster gateways and reach the gateway in the destination node's physical subnet. From there, the packets are delivered to the destination node within the subnet.

In the second situation, both the source and destination nodes are cluster gateways of different virtual clusters, but neither of these clusters is shared by both the source and destination's physical subnets. Here, routing packets from the source node to the destination node requires using one or more intermediate physical subnets. The ideal situation is that there exists a physical subnet that has cluster gateways for both the source and destination node's virtual clusters. This has been illustrated in Figure 2.14. In this situation, the packets use the source node's virtual cluster to reach the intermediate physical subnet. Within the subnet, packets move from the source's virtual cluster to an intermediate gateway of the destination's

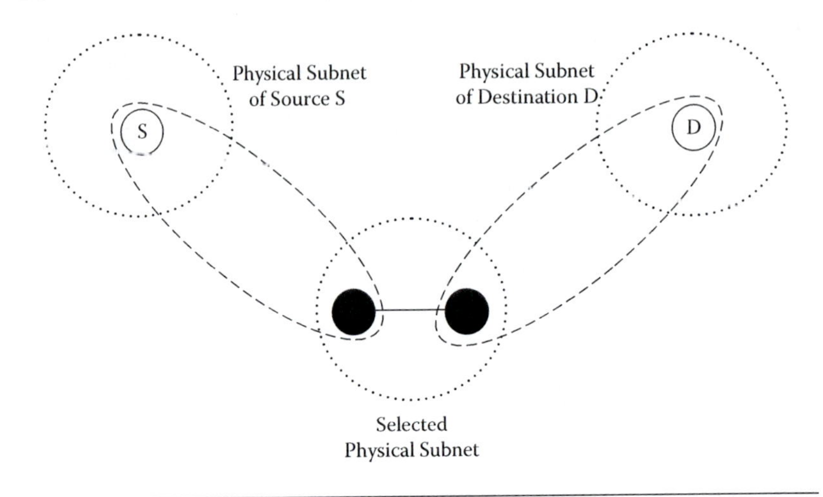

Figure 2.14 Long path routing using virtual clusters.

virtual cluster. From there, packets move along the destination node's virtual cluster to reach the destination node.

This is, however, the ideal situation where only one intermediate physical subnet is for reaching the destination. The concept can be easily extended to multiple hops of intermediate physical subnets before finding a subnet containing a cluster gateway for the destination node's virtual cluster.

2.5.2 The Hierarchical Clustering Model

Hierarchical clustering can also be termed multilevel clustering. In this model, the nodes of a wireless ad-hoc network are grouped together to form physical clusters. These clusters are further grouped together to form higher-level clusters, and so on. The independent wireless nodes that form the network can be viewed as Level-0 clusters. The physical clusters formed from these Level-0 clusters are called Level-1 clusters and are identical to the physical clusters defined under the flat clustering model. However, the architecture continues. Level-1 clusters are combined to form Level-2 clusters, and the process can be continued. The highest-level cluster (Level-N cluster) encompasses all the nodes that have joined the network.

Figure 2.15 illustrates the idea of hierarchical clustering. A Level-3 cluster has been designed that includes three Level-2 clusters, which in turn include seven Level-1 clusters altogether. The figure clearly

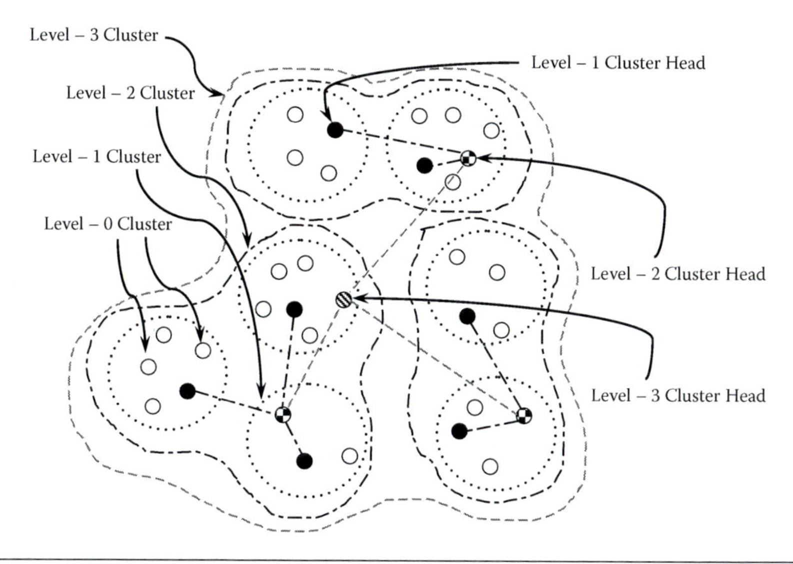

Figure 2.15 Symmetric hierarchical clustering model.

points out how the nodes are grouped to form multilevel clusters. Thus, it is easy to observe that the Level-3 cluster head is also a Level-0 cluster. However, there is an underlying symmetry in the figure. All Level-0 clusters are at the same level of hierarchy.

This may not always be the case. It is not necessary that only Level-K clusters combine to form Level-(K + 1) clusters. It may also be the case that a Level-(K + 1) cluster consists of some Level-K clusters, some Level-(K – 1) clusters, some Level-(K – 2) clusters, and so on. An example of this is shown in Figure 2.16.

Figure 2.16 shows an asymmetric hierarchical cluster that has four levels. However, the Level-3 cluster has two Level-2 clusters, one Level-1 cluster, and four Level-0 clusters making up the hierarchy. This is significant since the height of the routing tree no longer remains uniform and balanced.

There are two variations to the hierarchical model based on how packets are routed from a source node to a destination node. The rigid and flexible hierarchical models basically differ in how packets are routed through the network utilizing the cluster hierarchy.

2.5.2.1 Rigid Hierarchical Clustering Let us assume a symmetric hierarchical clustering model as shown in Figure 2.15. This implies that a Level-N cluster consists of only Level-(N – 1) clusters, and

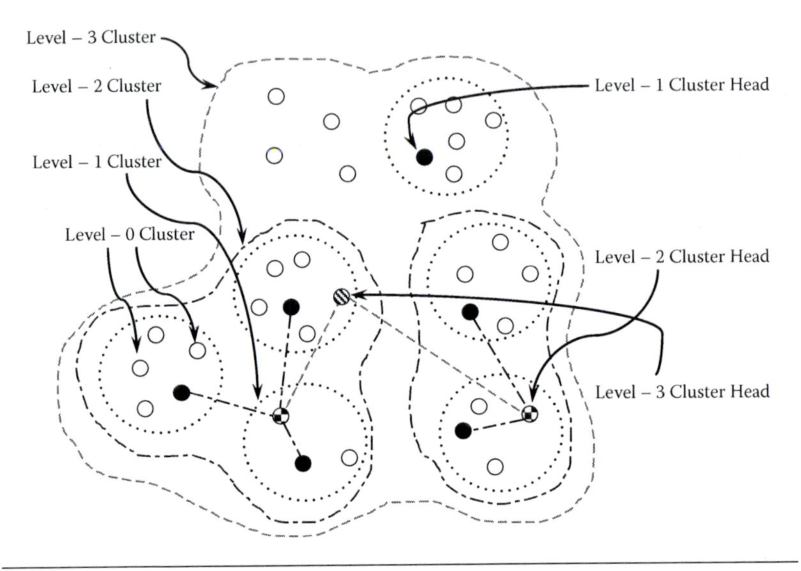

Figure 2.16 Asymmetric hierarchical clustering model.

Level-(N – 1) clusters contain only Level-(N – 2) clusters, and so on. The consequence is that all Level-0 clusters, including the source S and destination D, are at the same and lowest level of the hierarchy. If this is the case, the rigid hierarchical routing works as follows (Figure 2.17):

- Let the Level-K cluster be the lowest-level cluster that contains both the source node and the destination node.
- Since it is a symmetric hierarchy, there exists a node, say S_{K-1}, that is the source node's Level-(K – 1) cluster head and another node, say D_{K-1}, that is the destination node's Level-(K – 1) cluster head.
- Maintaining the symmetry, the S_{K-1} cluster contains S_{K-2}, S_{K-3}, …, S_1, which are the source node's Level-(K – 2), Level-(K – 3), …, Level-1 cluster heads, respectively. Let the source node be represented as S_0.
- Maintaining the symmetry, the D_{K-1} cluster also contains D_{K-2}, D_{K-3}, …, D_1, which are the destination node's Level-(K – 2), Level-(K – 3), …, Level-1 cluster heads, respectively. Let the destination node be represented as D_0.
- A packet from S_0 first travels up the hierarchy from S_0 through S_1, S_2, …, S_{K-2} and finally to S_{K-1}.
- The packet then traverses multiple Level-(K – 1) clusters and reaches the destination's Level-(K – 1) cluster head D_{K-1}.

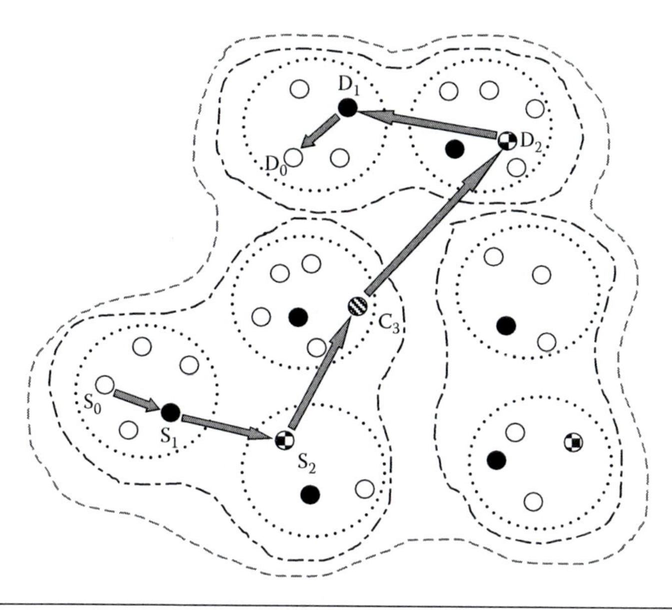

Figure 2.17 Rigid hierarchical routing.

- The packet again goes down the hierarchy from D_{K-1} through D_{K-2}, D_{K-3}, …, D_2 and finally to D_1. D_1 delivers the packet to the destination D_0.

2.5.2.2 Flexible Hierarchical Clustering Let us again assume a symmetric hierarchical clustering model, as shown in Figure 2.15. The flexible hierarchical routing works as follows:

- Let the Level-K cluster be the lowest-level cluster that contains both the source node S_0 and the destination node D_0.
- Since it is a symmetric hierarchy, there exists a node, say D_{K-1}, that is the destination node's Level-(K – 1) cluster head.
- Maintaining the symmetry, the D_{K-1} cluster contains D_{K-2}, D_{K-3}, …, D_1, which are the destination node's Level-(K – 2), Level-(K – 3), …, Level-1 cluster heads, respectively. Let the destination node be represented as D_0.
- A packet from S_0 reaches the destination's Level-(K – 1) cluster head, D_{K-1}, via the best possible route.
- From there the packet travels down the hierarchy from D_{K-1} through D_{K-2}, D_{K-3}, …, D_2 and finally to D_1. D_1 delivers the packet to the destination D_0.

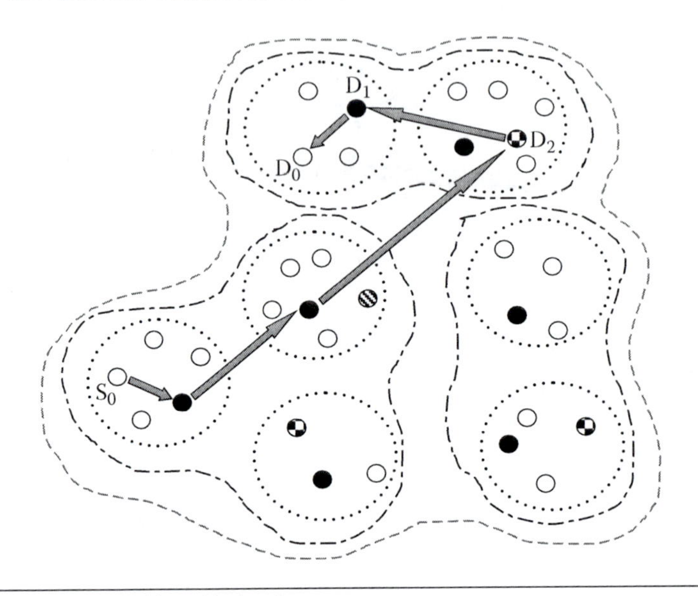

Figure 2.18 Flexible hierarchical routing.

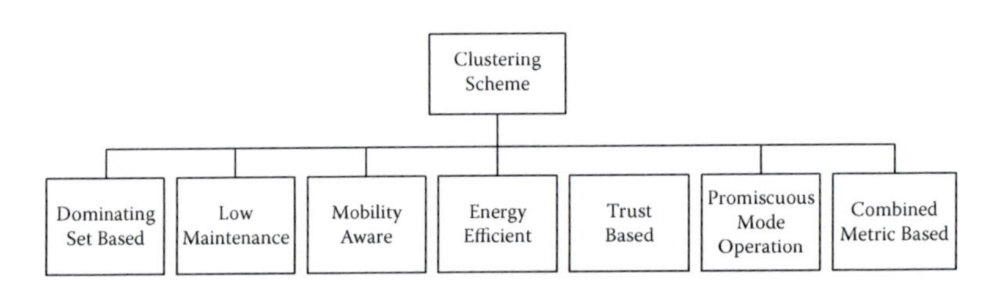

Figure 2.19 Taxonomy of clustering approaches.

Figure 2.18 illustrates the flexible hierarchical routing technique.

Clustering schemes in infrastructure-less networks can also be categorized, depending on the objective of the cluster formation. A classification of clustering techniques on multiple performance criteria is proposed in Figure 2.19.

2.5.3 Dominating Set (DS)-Based Cluster

Before going into detail on the DS-based cluster, we should gain knowledge about dominating set (DS). A dominating set is a set of all such nodes that have the responsibility of maintaining the routing table and monitoring the other member nodes in a cluster [32, 33]. If we consider a WAN as an undirected and unweighted graph G, a

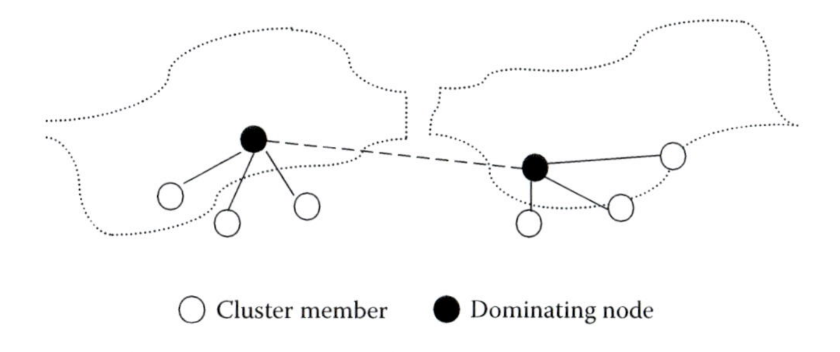

Figure 2.20 Connected dominating set.

subset D of G is said to be a dominating set if it includes all the monitoring nodes of the network.

In Figure 2.20, the dominating node has control over the dotted area around itself. The dominating nodes are connected themselves to reduce the routing overhead in WAN. In [32, 33] the authors have introduced the dominating set approach.

2.5.4 Low-Maintenance Clustering

In WAN cluster formation, cluster maintenance and cluster head selection need extra explicit message interchanges among nodes. From Chapter 1, we know that dynamic topology is the basic criterion of WAN. Frequent changes in network topology cause frequent cluster updating that increase overhead in the network and drastically decrease the network performance. Clustering in WAN consumes network bandwidth, which degrades the throughput of the network. To improve network performance, the communication overhead should be decreased. To reduce communication overhead, the following two parameters are taken into consideration:

1. Cluster restructuring: It refers to the re-formation of cluster. Whenever the topology changes, the cluster should be restructured with the nodes that newly join the network.
2. Cluster member reaffiliation: When a node joins a cluster in the network, the node has to make a new affiliation with the cluster to be a member of that new cluster.

To reduce restructuring and reaffiliation, a new role cluster guest is introduced in a cluster. Whenever a new node joins a cluster, it can join as a guest. The guest node can communicate with the cluster head by the help of another member node. This reduces restructuring of the cluster and reaffiliation also. In [34, 35], the authors have described a clustering algorithm where they introduce the concept of cluster guest.

2.5.5 *Mobility-Aware Clustering*

As has been described in the preceding section, the nodes in WAN are free to move anywhere in the network. So, mobility plays an important role in formation and maintenance of clusters in WAN. To make a stable cluster, all the nodes with similar mobility are grouped together, so that the connectivity among the nodes becomes tight enough. During cluster head selection the mobility of the node is taken into consideration. The node with minimum mobility is selected for the cluster head, so that the cluster becomes more stable and reclustering and reaffiliation of the cluster can be reduced. The algorithms described in [36, 37] have proposed mobility-aware clustering.

2.5.6 *Energy-Efficient Clustering*

Clustering is an approach in WAN that introduces a partially centralized and a partially distributed environment. In clustering, the nodes having similar frequency form a cluster. The nodes that initiate cluster formation declare themselves a cluster head. After formation of a cluster with a few nodes, the cluster head has to perform the following tasks:

1. Cluster maintenance
2. Member node monitoring
3. Routing table maintenance
4. Trust value computation
5. Communication among cluster members
6. Communication with cluster heads in different clusters

Each of these tasks consumes energy of the cluster head. So, to perform all these above-mentioned tasks, a cluster head should consist of more battery power. As we know, in WAN battery power is one of the most vital problems. To form an energy-efficient cluster we should set a limit on the size of the cluster. As the size of a cluster grows, more energy is consumed. The size of a cluster can be limited by introducing a restriction on the number of cluster members under a cluster head. To ensure the stability of a cluster, a cluster member having maximum battery power is chosen as a cluster head. The authors have described in [38–41] different approaches to create an energy-efficient stable cluster where the cluster head is equipped with sufficient energy and the load on the cluster head is restricted.

2.5.7 *Trust-Based Clustering*

In WAN the nodes work in cooperative fashion. Due to a short transmission range, multihop packet transfers are often observed in WAN. For reliable transmission of packets from the source to the destination node, the intermediate node should be trustworthy. So, we can say that trustworthiness is the basic requirement for each node in WAN. To ensure the trustworthiness of each member in a cluster, the cluster head periodically evaluates the trust value of each member in its own cluster. This trust value is periodically updated by the cluster head. During evaluation and updating of the trust value, other members of the cluster also provide their opinion for that member. During cluster head selection with the other metrics, like battery backup, maximum connectivity, and mobility, the trust value of a member node also plays an important role. The node with the highest trust value is chosen as the cluster head for a cluster. The nodes with the least trust value are marked as unreliable and are blacklisted. Blacklisted nodes are never chosen as intermediate nodes while a node discovers a route to the destination node.

2.5.8 *Promiscuous Mode Operation-Based Clustering*

In WAN, different clusters use different frequencies for communicating among the cluster members and with the cluster head. So, it is obvious that the member of one cluster cannot hear the member of another cluster. But, some nodes in WAN are placed in the gateway of two or more clusters. The gateway nodes can communicate with the members of more than one cluster. In a clustered network, when source and destination nodes belong to the same cluster, they can communicate directly through their cluster head. But, when the source and destination nodes are in different clusters, the cluster head at layer 1 should communicate with the cluster head at layer 2 to find out the route to the destination node in different clusters. If the cluster consists of a gateway node that has a direct link with the destination node in a different cluster, then that node can be used for routing the packets. Due to the presence of such a gateway node, the communication delay can be reduced, and thus the network performance can be improved.

2.5.9 Combined Metric-Based Clustering

In the previous subsections there was discussion of various parameters of clustering. By introducing each parameter individually, we can improve some specific behavior of WAN; i.e., if we focus only on mobility, we can make a stable cluster, but we cannot reduce communication overhead. Next, if we concentrate only on energy efficiency, we cannot ensure trustworthy communication. So, here we consider the combination of all metrics that can be used to improve the performance of WAN in totality. In combined metric-based clustering the metrics that are combined together are energy efficiency, mobility, and trustworthiness of each member in a cluster. While cluster head selection is done, these three parameters are considered. The cluster member having less mobility, maximum battery power, and the maximum trust value is selected as the cluster head. This approach is considered to be the best while we are working in a clustering environment. But in reality, implementation of this kind of clustering approach is difficult with respect to mobility calculation, trust value computation, and battery power computation. The authors have proposed an algorithm in [42] where they have used combined metric-based clustering.

2.6 Wireless Mesh Networks (WMNs)

Although mobility of nodes was removed and a certain infrastructure was established for sensor networks, WANs remained vulnerable to security threats. Researchers realized that mobility is a feature that cannot be compromised, as it provides tremendous flexibility to end users. Yet, retaining an infrastructure would definitely be helpful. All these underlying observations led to the conclusion that a different type of network must be designed that incorporates both the mobility of clients and a basic infrastructure. This was the inception of wireless mesh networks.

Wireless mesh networks (WMNs) consist of mesh routers and mesh clients, where mesh routers have minimal mobility and form the backbone of WMNs. They provide network access for both mesh and conventional clients. The integration of WMNs with other networks, such as the Internet, cellular, IEEE 802.11, IEEE 802.15, IEEE 802.16, sensor networks, etc., can be accomplished through the gateway and bridging functions in the mesh routers. Mesh clients can be either stationary or mobile, and can form a client mesh network among themselves and with mesh routers. WMNs have evolved to eliminate the limitations and to significantly improve the performance of ad-hoc networks, wireless local area networks (WLANs), wireless personal area networks (WPANs), and wireless metropolitan area networks (WMANs). They are undergoing rapid progress and inspiring numerous deployments. WMNs will deliver wireless services for a large variety of applications in personal, local, campus, and metropolitan areas [31].

2.6.1 Hybrid Architecture of Wireless Mesh Networks

Infrastructure/backbone of WMNs: WMNs include mesh routers forming an infrastructure for clients that connect to them. The WMN infrastructure/backbone can be built using various types of radio technologies. The most frequently used technologies are the IEEE 802.11 technologies. The mesh routers form a mesh of self-configuring, self-healing links among themselves. With gateway functionality, mesh routers can be connected to the Internet. This approach, also referred to as infrastructure meshing, provides a backbone for conventional clients and enables integration of WMNs with existing wireless networks, through gateway/bridge functionalities in mesh routers.

For conventional clients with the same radio technologies as mesh routers, they can directly communicate with mesh routers. If different radio technologies are used, clients must communicate with the base stations that have Ethernet connections to mesh routers. Typically, two types of radios are used in the routers, i.e., for backbone communication and for user communication, respectively. The mesh backbone communication can be established using long-range communication techniques, including directional antennas.

Client WMNs: Client meshing provides peer-to-peer networks among client devices. In this type of architecture, client nodes constitute the actual network to perform routing and configuration functionalities as well as providing end user applications to customers. Hence, a mesh router is not required for these types of networks. In client WMNs, a packet destined to a node in the network hops through multiple nodes to reach the destination. Client WMNs are usually formed using one type of radio on devices. Moreover, the requirements on end user devices are increased when compared to infrastructure meshing, since in client WMNs, the end users must perform additional functions, such as routing and self-configuration.

Hybrid WMNs: This architecture is the combination of infrastructure and client meshing, as shown in Figure 2.21. Mesh clients can access the network through mesh routers, as well as directly meshing with other mesh clients. While the infrastructure provides connectivity to other networks such as the Internet, Wi-Fi, WiMAX, cellular, and sensor networks, the routing capabilities of clients provide improved connectivity and coverage inside the WMN. The hybrid architecture will be the most applicable case, in our opinion.

What's so good about mesh networks?

- The redundancy and self-healing capabilities provide for less downtime, with messages continuing to be delivered even when paths are blocked or broken.
- The self-configuring, self-tuning, self-healing, and self-monitoring capabilities of mesh can help to reduce the management burden for system administrators.
- Advanced mesh networking protocols coordinate the network so that nodes can go into sleep mode while inactive and then synchronize quickly for sending, receiving, and forwarding messages. This ability provides greatly extended battery life.

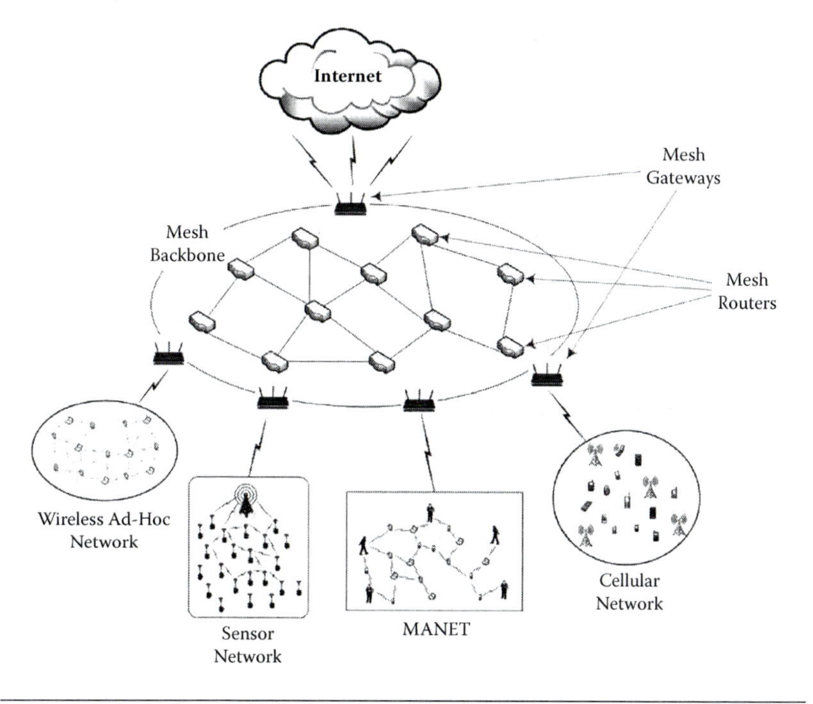

Figure 2.21 Wireless mesh network.

- A mesh network can be deliberately overprovisioned simply by adding extra devices, so that each device has two or more paths for sending data. This is a much simpler and less expensive way of obtaining redundancy than is possible in most other types of networks.
- Compared to the cost of point-to-point copper wiring and conduit required for traditional wired networks, wireless mesh networks are typically much less expensive. The self-management capabilities of mesh networks can also help to make installation less expensive than for traditional networks.

2.7 Conclusions

As networks have evolved over time, so have their vulnerabilities. WMNs are the latest trend in network architecture deployment. However, this architecture also has its drawbacks and limitations. Moreover, with the advent of WMNs, MANETs, sensor, and other ad-hoc networks have not become obsolete. Each type of infrastructureless wireless network has its own application area. Attackers are continuously exploiting the features of these networks and inventing ingenious attacks that are becoming increasingly difficult to detect. The next chapters of this book are dedicated to detecting and isolating these attacks from wireless ad-hoc infrastructure-less networks.

References

1. Andrew S. Tanenbaum, *Computer Networks*, 4th ed., Pearson Education/Prentice Hall, India, 2002, pp. 1–891.
2. Scott Jordan, Do Wireless Networks Merit Different Net Neutrality Than Wired Networks? *TPRC*, October 2010.
3. Charles E. Perkins, Cluster-Based Networks, in *Ad hoc Networking*, chap. 4, Pearson Education and Dorling Kindersley Publishing, India, 2008.
4. L. Zhou and Z.J. Haas, Securing Ad hoc Networks, *IEEE Network Magazine*, November/December 1999.
5. N. Asoka and P. Ginzboorg, Key-Agreement in Ad-Hoc Networks, *Computer Communications*, 23(17) November 2000, pp. 1627–1637.
6. Dan Zhou, *Security Issues in Ad hoc Networks*, Florida Atlantic University, 2003.
7. Amitabh Mishra and Ketan M. Nadkami, The Electrical Engineering Handbook Series, in *The Handbook of Ad hoc Wireless Networks*, CRC Press, Boca Raton, FL, 2003, pp. 499–549.
8. P. Papadimitratos and Z.J. Haas, Secure Routing for Mobile Ad hoc Networks, at Proceedings of the SCS Communication Networks and Distributed System Modeling and Simulation Conference (CNDS 2002), 2002.
9. I.F. Akyildiz, W. Su, Y. Sankarasubramaniam, and E. Cayirci, A Survey on Sensor Networks, *IEEE Communications Magazine*, 40(8), 102–114, 2002.
10. A. Perrig, R. Szewczyk, J.D. Tygar, V. Wen, and D.E. Culler, Spins: Security Protocols for Sensor Networks, *Wireless Networking*, 8(5), 521–534, 2002.
11. J. Deng, R. Han, and S. Mishra, *INSENS: Intrusion-Tolerant Routing in Wireless Sensor Networks*, Technical Report CU-CS-939-02, Department of Computer Science, University of Colorado at Boulder, 2002.
12. B. Karp and H.T. Kung, GPSR: Greedy Perimeter Stateless Routing for Wireless Networks, in *Proceedings of the 6th Annual International Conference on Mobile Computing and Networking*, ACM Press, New York, 2000, pp. 243–254.
13. S. Tanachaiwiwat, P. Dave, R. Bhindwale, and A. Helmy, Poster Abstract Secure Locations: Routing on Trust and Isolating Compromised Sensors in Location Aware Sensor Networks, in *Proceedings of the 1st International Conference on Embedded Networked Sensor Systems*, ACM Press, New York, 2003, pp. 324–325.

14. D. Estrin, R. Govindan, J.S. Heidemann, and S. Kumar, Next Century Challenges: Scalable Coordination in Sensor Networks, *Mobile Computing and Networking*, 1999, pp. 263–270.
15. L. Hu and D. Evans, Secure aggregation for wireless networks, in *Proceedings of the Workshop on Security and Assurance in Ad Hoc Networks*, January 2003, Orlando, FL.
16. S. Madden, M.J. Franklin, J.M. Hellerstein, and W. Hong, *Tag: A Tiny Aggregation Service for ad-Hoc Sensor Networks, SIGOPS Operating Systems Revision*, 36(SI), 131–146, 2002.
17. B. Przydatek, D. Song, and A. Perrig, SIA: Secure Information Aggregation in Sensor Networks, in *Proceedings of the 1st International Conference on Embedded Networked Sensor Systems (SenSys '03)*, November 2003, pp. 255–265.
18. N. Shrivastava, C. Buragohain, D. Agrawal, and S. Suri, Medians and Beyond: New Aggregation Techniques for Sensor Networks, in *SenSys'04: Proceedings of the 2nd International Conference on Embedded Networked Sensor Systems*, ACM Press, New York, 2004, pp. 239–249.
19. F. Ye, H. Luo, S. Lu, and L. Zhang, Statistical En-Route Detection and Filtering of Injected False Data in Sensor Networks, in *IEEE INFOCOM 2004*.
20. A.R. Beresford and F. Stajano, Location Privacy in Pervasive Computing, *IEEE Pervasive Computing*, 2(1), 46–55, 2003.
21. T. Kaya, G. Lin, G. Noubir, and A. Yilmaz, Secure Multicast Groups on Ad hoc Networks, in *Proceedings of the 1st ACM Workshop on Security of Ad hoc and Sensor Networks (SASN'03)*, ACM Press, New York, 2003, pp. 94–102.
22. S. Rafaeli and D. Hutchison, A Survey of Key Management for Secure Group Communication, *ACM Computer Survey*, 35(3), 309–329, 2003.
23. S. Ganeriwal and M. Srivastava, Reputation-Based Framework for High Integrity Sensor Networks, at Proceedings of the 2nd ACM workshop on Security of Ad hoc and Sensor Networks, Washington, DC, 2004.
24. Z. Liang and W. Shi, *Analysis of Recommendations on Trust Inference in the Open Environment*, Technical Report MIST-TR-2005-002, Department of Computer Science, Wayne State University, Detroit, MI, February 2005.
25. Z. Liang and W. Shi, Enforcing Cooperative Resource Sharing in Untrusted Peer-to-Peer Environment, *ACM Journal of Mobile Networks and Applications (MONET)*, 10(6), 771–783, 2005.
26. Z. Liang and W. Shi, PET: A Personalized Trust Model with Reputation and Risk Evaluation for P2P Resource Sharing, at Proceedings of the HICSS-38, Hilton Waikoloa Village Big Island, HI, January 2005.
27. K. Ren, T. Li, Z. Wan, F. Bao, R.H. Deng, and K. Kim, Highly Reliable Trust Establishment Scheme in Ad hoc Networks, *Computer Networks: The International Journal of Computer and Telecommunications Networking*, 45, 687–699, 2004.

28. S. Tanachaiwiwat, P. Dave, R. Bhindwale, and A. Helmy, Location-Centric Isolation of Misbehavior and Trust Routing in Energy-Constrained Sensor Networks, at IEEE Workshop on Energy-Efficient Wireless Communications and Networks (EWCN04), in conjunction with IEEE IPCCC, April 2004.
29. Z. Yan, P. Zhang, and T. Virtanen, Trust Evaluation Based Security Solution in Ad hoc Networks, at NordSec 2003, Proceedings of the Seventh Nordic Workshop on Secure IT Systems, 2003.
30. H. Zhu, F. Bao, R.H. Deng, and K. Kim, Computing of Trust in Wireless Networks, at Proceedings of the 60th IEEE Vehicular Technology Conference, Los Angeles, CA, September 2004.
31. I.F. Akyildiz, X. Wang, and W. Wang, Wireless Mesh Networks: A Survey, *Computer Networks and ISDN Systems*, 47(4), March 2005, pp. 445–487.
32. J. Wu and H.L. Li, On Calculating Connected Dominating Set for Efficient Routing in Ad hoc Wireless Networks, in *Proceedings of the 3rd International Workshop on Discrete Algorithms and Methods for Mobile Computing and Communications*, 1999, pp. 7–14.
33. Y.-Z.P. Chen and A.L. Liestman, Approximating Minimum Size Weakly-Connected Dominating Sets for Clustering Mobile Ad hoc Networks, in *Proceedings of the 3rd ACM International Symposium on Mobile Ad hoc Networks and Computing*, June 2002, pp. 165–72.
34. C.-C. Chiang et al., Routing in Clustered Multihop, MobileWireless Networks with Fading Channel, in *Proceedings of IEEESICON'97*, 1997.
35. J.Y. Yu and P.H.J. Chong, 3hBAC (3-Hop between Adjacent Clusterheads): A Novel Non-Overlapping Clustering Algorithm for Mobile Ad hoc Networks, in *Proceedings of IEEE Pacrim'03*, August 2003, vol. 1, pp. 318–21.
36. P. Basu, N. Khan, and T.D.C. Little, A Mobility Based Metric for Clustering in Mobile Ad hoc Networks, in *Proceedings of IEEE ICDCSW'01*, April 2001, pp. 413–418.
37. A.B. McDonald and T.F. Znati, Design and Performance of a Distributed Dynamic Clustering Algorithm for Ad-Hoc Networks, in *Proceedings of the 34th Annual Simulation Symposium*, April 2001, pp. 27–35.
38. J. Wu et al., On Calculating Power-Aware Connected Dominating Sets for Efficient Routing in Ad hoc Wireless Networks, *Journal of Communications and Networks*, 4(1), 59–70, 2002.
39. J.-H. Ryu, S. Song, and D.-H. Cho, New Clustering Schemes for Energy Conservation in Two-Tiered Mobile Ad-Hoc Networks, in *Proceedings of IEEE ICC'01*, June 2001, vo1. 3, pp. 862–866.
40. T. Ohta, S. Inoue, and Y. Kakuda, An Adaptive Multihop Clustering Scheme for Highly Mobile Ad hoc Networks, at Proceedings of 6th ISADS'03, April 2003.
41. A.D. Amis and R. Prakash, Load-Balancing Clusters in Wireless Ad hoc Networks, in *Proceedings of 3rd IEEE ASSET'00*, March 2000, pp. 25–32.
42. M. Chatterjee, S.K. Das, and D. Turgut, An On-Demand Weighted Clustering Algorithm (WCA) for Ad hoc Networks, in *Proceedings of IEEE Globecom'00*, 2000, pp. 1697–1701.

3

Routing for Ad-Hoc Networks

DEBDUTTA BARMAN ROY
AND RITUPARNA CHAKI

Contents

3.1 Introduction

The infrastructure-less nature of wireless ad-hoc networks (WANs) has already been discussed. The nodes are capable of forming a network on an anywhere, anytime basis. The communication between the nodes takes place in a distributed manner. The establishment of communication channels between a pair of source and destination nodes becomes difficult due to the lack of infrastructure and absence of centralized authority. A closer look needs to be taken at this problem.

The process of transferring information from a source to a destination within a network is known as routing. During this process, at least one intermediate node within the network is encountered, barring the case of the next hop node being the destination node. The routing concept basically involves two activities: (1) determining optimal routing paths and (2) transferring the information in the form of packets through a network.

To compute the most optimum path from the source to the destination node, routing protocols use several metrics. After finding out the optimum path from the source to the destination, routing table information is updated by each node that lies in the path from the source to the destination node. Different routing algorithms maintain the routing table in their own way.

As has been said, WANs have many key characteristics [36] that distinguish them from wired networks. The key characteristics that make traditional routing protocols unsuitable for WAN are mentioned here:

1. *Dynamic topology*: In WAN a rapid change in topology is caused by the random movement of nodes in the network. The nodes can join or leave the network any time. To route a packet in WAN, multihop routing is required due to the short transmission range.
2. *Variable capacity links*: Compared to wired links, the wireless links have lower link capacity. Wireless links can be of two types: unidirectional and bidirectional.
3. *Congestion*: Due to limited link capacity, congestion is a big problem for WAN. Compared to wired links, in wireless links the link capacity is readily reachable.
4. *Energy-constrained mobile nodes*: In WAN the nodes usually operate on batteries. Sometimes, to save the batteries, the nodes may go into sleep mode. That may cause a delay in response. So, in WAN the nodes operate in the most optimized way.
5. *Weakened physical security*: WAN is much more vulnerable to physical security threats than wired networks.

From the previous discussion, we came to know that WANs need a kind of routing protocol that can work under distributed, decentralized, and infrastructure-less environments. Since the beginning of the Defense Advance Research Project Agency packet radio network in the early 1970s, numerous routing protocols have been developed for WAN. These routing protocols are broadly classified according to their process of route discovery. To get a clear idea about the classification of routing protocols, the taxonomy is given in Figure 3.1. Due to the endless research in this domain, it is not possible to include all the proposed routing protocols in a single figure.

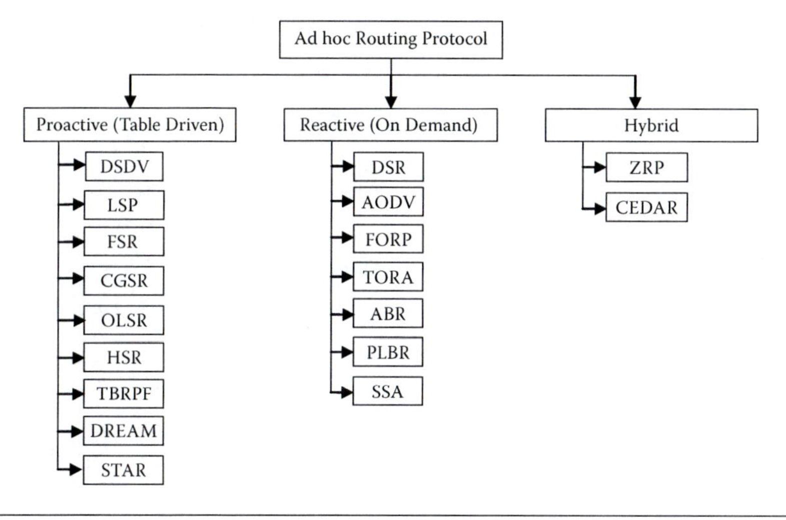

Figure 3.1 Taxonomy of routing protocol.

3.2 Routing for WAN

This section deals with different routing protocols in WANs. Depending on the process of route discovery, routing techniques are broadly classified into three types: proactive, reactive, and hybrid routing protocols. Detailed studies of all these protocols are given in the following subsections.

3.2.1 Proactive Routing

Each node in the network maintains routing information for every other node in the network in a routing table. The routing table is periodically updated as the network topology changes. Many of these routing protocols come from the link state routing [5]. There exist some differences between the protocols that come under this category, depending on the routing information being updated in each routing table. The proactive protocols are not suitable for larger networks, as they need to maintain node entries for each and every node in the routing table of every node. This causes more overhead in the routing table, leading to consumption of more bandwidth. Some of the renowned proactive routing protocols are described here.

The destination sequenced distance vector (DSDV) routing protocol [7, 8, 41] is an enhancement of the Bellman-Ford algorithm.

Updating the routing table and sequence number leads to preventing problems like loops and count-to-infinity. In this mechanism, routes to all destinations are readily available at every node at all times. The tables are exchanged between neighbors at regular intervals to keep an up-to-date view of the network. Neighbor nodes use missing transmissions to detect broken links in the topology. When a broken link is found, it is assigned a metric value of infinity, and the node that detected the broken link broadcasts an update packet to inform others that another link has been chosen.

The link state protocol (LSP) [9] is based on link state algorithm, and it is proactive in nature. It reduces the size of messages during transmission, which controls flooding. It uses an optimal number of hops in the entire network.

The fisheye state routing (FSR) protocol [10] is based on link state routing. FSR was designed to reduce message overhead in dynamic environments. Link state routing information broadcasts the updated information throughout the network, whereas in FSR routing, information is disseminated. In this, a node rapidly shares information with its nearest neighborhoods and less frequently with distant nodes. Thus, it alleviates the problem of message overhead, but it increases the bandwidth issue when node density increases.

The cluster head gateway search routing (CGSR) protocol [11] is designed to provide effective membership and traffic management. It is based on the distance vector routing protocol. In this protocol, the whole network can be partitioned into clusters. Each cluster maintains one cluster head and at least one gateway node. By using the concept of cluster, it reduces the size of the table compared to the distance vector protocol. But the maintenance of cluster structure is very difficult in CGSR.

The optimized link state routing (OLSR) protocol is an enhancement [1] of LSR and follows the concept of multipoint relays. Nodes regularly broadcast beacon messages to their one-hop neighbors, which include the list of neighbors to which a link exists.

The distance routing effect algorithm for mobility (DREAM) is a proactive [12] routing protocol that follows the concept of directional flooding to forward data packets. Thus, there will be multiple copies of each packet at the same time. This increases the probability of using the optimal path; however, it decreases its scalability in large-scale networks.

The topology broadcast based on reverse-path forwarding (TBRPF) protocol is another link state, proactive routing protocol for WANs [39, 40]. Each router running TBRPF computes a source tree to all reachable destinations based on partial topology information stored locally. The source tree is also known as the shortest path tree [39]. To reduce overhead, routers in TBRPF only broadcast part of their source tree to neighbors. The partial source tree is called the reportable tree. The main idea of sharing reportable trees with neighbors comes from the partial tree-sharing protocol (PTSP) described in [38]. Basically, in the local copy of network topology, a link's cost is equal to the actual value if this link is in the shortest path tree. Otherwise, the cost is equal to or greater than the real value. The procedure to generate a reportable tree at a router is as follows. Links that are in this router's shortest path tree are checked. If such a link is estimated to be in the neighbors' shortest path trees, it is added to the reportable tree. Note that the estimated results may not be correct, but they do include the correct link costs. TBRPF is said to work better in dense networks [40]. TBRPF has two modules, the neighbor discovery (TND) module and the topology discovery and route computation module. "The key feature of TND is that it uses 'differential' HELLO messages which report only changes in the status of neighbors" [40]. This reduces the size of HELLO packets used in this module. The HELLO packet in TBRPF may contain three lists of router IDs. They are in three different formats: neighbor request, neighbor reply, and neighbor lost. The neighbor request list includes the IDs of new neighbors whose HELLO messages are heard for the first time. This implies that the links to those neighbors are currently one-way links. Note that the neighbor request list is always included in HELLO packets, even if it is empty. The other two lists may not be included if they are empty. HELLO packets are sequenced by senders. The TND module is responsible for discovering any new neighbors and detecting the loss of any neighbors. After continuously hearing HELLO packets a certain number of times, usually twice, a router responds by sending HELLO packets with a neighbor request set in each of its next *NBR HOLD COUNT* (typically three) HELLO messages, or until a neighbor reply message is received from the new neighbor. This avoids short-lived links. When a router receives a neighbor's reply message, it declares a bidirectional link by sending *NBR HOLD COUNT* HELLO messages, including the received neighbor reply message. In the case of missing HELLO packets for *NBR HOLD COUNT* times, a router sends *NBR HOLD COUNT* neighbor lost messages to that neighbor. A neighbor is declared lost if no reply comes from that neighbor. TBRPF has two modes of operation: full topology (FT) mode and partial topology (PT) mode. Bellur and Ogier describe the FT mode [37], which with Templin and Lewis they later modify to be the PT [40]. In PT each node receives only a PT graph that is just enough to construct the shortest path to all nodes. The PT mode is used in simulation experiments for this study. Using a modified version of Dijkstra's algorithm, each node forms a shortest path source tree to all reachable nodes. TBRPF reports only partial source trees in its link state updates instead of the entire neighbor link costs. The PT graph, which is called the reportable subtree (RT), is a subgraph of the source tree at the node. Unlike other proactive routing protocols, TBRPF does not periodically broadcast FT updates. Instead, it broadcasts a combination of periodic and differential updates of the RT. The changes to RT since the last full RT update are broadcast using differential updates; differential updates occur more periodically. TBRPF nodes also broadcast the full RT once every few seconds to allow newly detected nodes to construct the FT. The use of RT and differential updates minimizes control packet overhead. The nodes forward the topology updates along the reverse-path tree for any source. Reverse-path forwarding is achieved by using the information obtained by TBRPF operation. The TBRPF node forwards if the originating source node is a member of the reportable node (RN) set computed by TBRPF. The reportable node set is a set that includes all neighbors *j* of a node *i* if node *i* determines that any of its neighbors might use node *i* as the next hop in the shortest path to node *j*. The algorithm to form a RT is given in TBRPF specifications. The RT at a node contains all the local links to its one-hop neighbors and the subtrees of the source tree rooted at neighbors in the RN set [40].

3.2.2 Reactive Routing

In this routing protocol, if a node wants to send a packet to another node, then this protocol searches for the route in an on-demand manner and establishes the connection in order to transmit and receive the packet [16]. The route discovery usually occurs by flooding the route request packets throughout the network. A few well-known reactive routing protocols are described in this subsection.

The dynamic source routing (DSR) protocol [3] is an on-demand protocol that establishes a route by flooding route request packets in the network. The destination node responds by sending a route reply packet back to the source, along with the route information. While the DSR protocol eliminates the need to periodically flood the network with the table update messages, the major limitation of this protocol is that the route maintenance mechanism cannot repair a broken link locally. Moreover, the route identified by DSR need not necessarily be the shortest one in between the given pair of source and destination nodes.

The ad-hoc on-demand distance vector (AODV) [5] routing protocol is another example of an on-demand routing algorithm. AODV assumes symmetric links between neighboring nodes. It further assumes that neighboring nodes can detect each other's broadcasts. The primary difference between AODV and other on-demand routing protocols is that AODV uses a destination sequence number to determine the optimal path to a destination. While AODV offers loop-free routes even while repairing broken links and is also scalable to a large population of nodes, one of the disadvantages is that the intermediate nodes may have stale entries, and thus could lead to inconsistent routes. Thus, it is found that all the proactive routing algorithms that are used in WAN suffer from some severe common drawbacks [6]. Besides storing the topological information at all the hosts, these algorithms consume power and system resources, including bandwidth for repetitive computation of possible routing paths, with often a very poor usage ratio of the paths computed thus. Besides these, many of the proactive routing algorithms create long-lived loops and take too long to converge. On the other hand, for reactive topologies, the routing overhead in terms of connection setup delay increases rapidly when the number of sources increases. This may be attributed to the fact that in an on-demand approach, the nodes do not store any route tables.

The flow-oriented routing protocol (FORP) [1–3] is an on-demand routing protocol based on the pure flooding mechanism. Moreover, it maintains a prediction-based multihop handoff mechanism. This attempt is used to reduce the effect of communication failure. Its route request process is the same as that of DSR. Additionally, each node that receives *flow_req* calculates link expiration time (LET). Due to this, the destination could easily know when a route is about to expire. If it expires, the destination node generates a handoff message and propagates it by flooding. When a source node receives this message, it can easily identify the optimal path to handoff. Then the source sends a *flow_setup* message along the newly chosen route. This protocol generates a scalability problem in large networks.

The temporally ordered routing algorithm (TORA) [1, 3, 4] is a distributed, dynamic, and multihop routing protocol. It is based on the directed link reversal algorithms. This protocol is better than FORP in terms of efficiency, adaptability, and scalability for large, dense mobile networks. This protocol is designed to minimize reaction to topological changes at a very low rate.

The associativity-based routing (ABR) [1–4] is based on source-initiated on-demand routing. It is also known as a distributed long-lived routing protocol for ad-hoc networks. Due to this feature, this protocol results in higher attainable throughput. This protocol deals with fault tolerance when the base station fails.

The preferred link-based routing (PLBR) protocol minimizes control overhead by using a subset of the preferred list. Selection of this list can be based on degree of node.

The signal stability-based adaptive routing (SSA) protocol [2] is an on-demand routing protocol that considers signal stability a prime factor. In this signal, strength is used to rectify a link as stable or unstable. SSA provides more stable routes than shortest path routing protocols such as DSR and AODV. This protocol provides good link stability compared to other protocols. A comparative summary of the discussion for on-demand and table-driven routing is presented in Table 3.1.

Table 3.1 Comparative Study of On-Demand and Table-Driven Routing

PARAMETER	ON-DEMAND	TABLE-DRIVEN
Availability of routing information	Available when needed	Always available regardless of need
Routing philosophy	Flat	Mostly flat
Periodic route update	Not required	Required
Mobility handling	Use localized route discovery	Inform other nodes to achieve a consistent routing table
Congestion	Less congestion	More congestion due to periodic update of route
Delay	More end-to-end delay, because each time the route should be discovered first	Less end-to-end delay, as the route is already created
Routing attack	Less prone to routing attack as route is created on demand	More prone to routing attack

3.2.3 Hybrid Routing

After a detailed study of proactive and reactive routing protocols, the merits and demerits of both approaches are apparent. In this subsection, another approach of routing is discussed that includes only the merits of both aforesaid routing protocol strategies. Some of the renowned hybrid routing protocols are described here.

The zone routing protocol (ZRP) is known as a hybrid zone routing protocol [10, 14, 15]. Such protocols are suitable for large area networks by adjusting the transmission range of nodes. The configuration of the routing protocol is based on both proactive and reactive routing protocols. The core extraction distributed ad-hoc routing (CEDAR) protocol supports quality-of-service [13–16, 31] reliable mechanisms and is based on extracting core nodes in the network.

Defrawy and Tsudik have proposed anonymous location-aided routing in suspicious MANETs (ALARM). It is a link state routing protocol. It provides both security and privacy features, including node authentication, data integrity, and anonymity. Each node is equipped with a GPS-like device for accurate location information. Each node broadcasts a location announcement message (LAM), besides other details. Each LAM is flooded throughout the WAN, which eventually creates an overhead when the network size is large [27]. The overhead due to extra control signals was a serious problem. In their attempt to reduce this overhead, the researchers came up with a zone-based partitioning concept. The zone-based hierarchical link state (ZHLS) routing algorithm by Joa-Ng and Lu [28] and the segment by segment routing (SSR) protocol by Cao [29] partition the routing area into square regions, whereas VCR [30], proposed by Sivavakeesar and Pavlou, partitions it into circular-shaped regions. All of them organize the nodes within the same region into the same cluster. The difference is that ZHLS does not select a cluster head, whereas VCR and SSR do. The size of a geographic region is also different in different protocols. In adaptive cell relay (ACR) [31] by Du and Wu the routing area is first distributed into cells. A packet is routed along the cell chain that joins the source cell to the destination cell; the cell is small so that one node can directly communicate with other nodes in its neighboring cells. However, ZHLS, VCR, and SSR use large partitions.

In ZHLS, each node maintains an intrazone routing table for nodes within its zone. In VCR, each circular region constitutes a virtual cluster, and the node that is stable and close to the region center is elected as the cluster head. The radius of each virtual cluster is the same and is set to be *k* hops. In SSR [29], the cluster head is elected according to the same rule as in VCR [30]. However, SSR maintains the *k*-hop routing tables only at each cluster head. All these techniques involved the overhead of frequent update of the cluster head.

In zone and link expiry-based routing protocol (ZLERP), the stability of the link is based on the signal strength received at a periodic time interval by a node that is on the periphery of another node's zone. Variation in signal strength depends on many factors, such as distance between nodes, angles between nodes, obstacles, blocked regions, noise, interference, etc. ZLERP considers two main factors, distance between nodes and blocked terrains, in [43] by Manvi, Kakkasageri, Paliwal and Patil. However, the remaining unconsidered factors might often prove to be strong enough to hamper the throughput.

As both the cluster-based and signal-based approaches incurred high overhead, researchers focused on identification of exact location of nodes. Location-aided routing (LAR) [32, 33] utilizes the location information for routing, somewhat similar to zone-based routings [34, 35]. LAR assumes the availability of the geographical position information of nodes necessary for routing. In LAR or location-based directional route discovery (LDRD), a request zone is defined depending on the positions of the source and destination and also on the average node mobility in the network. The goal of location-aided routing is to reduce the routing overhead by the use of location information.

3.3 Security Issue in WAN

The previous section includes a discussion about the different kinds of routing techniques. This section looks at a special group of routing protocols that focus on secure message routing. The routing protocols discussed above do not consider secure transmission of information from source to destination node in WAN. This section includes a detailed discussion on secure data transmission over WAN. Managing the security of information in WAN consists of a two-way approach. One is the use of an intrusion detection system (IDS) to save the system from malicious attacks. The second part consists of securing the message itself by using special measures. Until now, the IDS-based approaches have been discussed. In this section, some preexisting secure routing algorithms for mobile ad-hoc networks are described. These routing protocols have been categorized according to the security approaches into a cryptographic approach, trust-based approach, and message authentication. Let's start our discussion with the most popular, the cryptographic approach.

3.3.1 Cryptographic Approach

In the cryptographic approach the source node and the destination node in WAN make a secure route between themselves by using a private key and public key pair. While broadcasting RREQ packets, the source node encrypts them by using its private key. The nodes in WAN that can hear this RREQ try to decrypt the RREQ with their own public key and the authorized key that has been authenticated by a central authority. The authenticated node only can access this RREQ, but it can actively take part in formation of the route. Some existing cryptographic approaches are described in this subsection.

Kumaran et al. [42] have proposed building secure routing out of an incomplete set of security associations (BISS) [19]; the sender and the receiver can establish a secure route, even if, prior to the route discovery, only the receiver has security associations established with all the nodes on the chosen route. Thus, the receiver will authenticate route nodes directly through security associations. The sender, however, will authenticate directly the nodes on the route with which it has security associations, and indirectly (by exchange of certificates) the node with which it does not have security associations. The operation of BISS ROUTE REQUEST relies on mechanisms similar to direct route authentication protocols. When an initiator sends a ROUTE REQUEST, it signs the request with its private key and includes its public key infrastructure (PKI) in the request, along with a certificate signed by the central authority binding its ID with PKI. This enables each node on the path to authenticate the initiator of the ROUTE REQUEST. The ROUTE REQUEST message contains the ID of the target node. The node that receives this ROUTE REQUEST authenticates the initiator (by verifying the signature on the message) and tries to authenticate the target directly through security associations that it has. Only if a node can successfully authenticate both the initiator and the target will the node broadcast the message further.

Xue and Nahrstedt proposed design and evaluation of ARIADNE, a new ad-hoc network routing protocol that provides security against one compromised node and arbitrary active attackers, and relies only on efficient symmetric cryptography [24]. ARIADNE operates on demand, dynamically discovering routes between nodes only as needed; the design is based on the basic operation of the DSR protocol. Rather than generously applying cryptography to an existing protocol to achieve security, however, the author carefully redesigned each protocol message and its processing. The security mechanisms the authors have designed are highly efficient and general, so that they should be applicable to securing a wide variety of routing protocols. This article presents the timed efficient stream loss-tolerant authentication (TESLA) broadcast authentication protocol, an efficient protocol with low communication and computation overhead, which scales to large numbers of receivers and tolerates packet loss. TESLA is based on loose time synchronization between the sender and the receivers. The authors have used low computation overhead in terms of generation and verification of authentication information as a metric for measuring the performance of TESLA. Another metric that is suggested by the author is low communication overhead, which can be achieved if there is limited buffering at both the sender and the receiver end.

Kim and Sudik [17] focus on securing the route discovery process in DSR. Their goal is to explore a range of suitable cryptographic techniques with varying flavors of security, efficiency, and robustness. The ARIADNE approach (with TESLA), while very efficient, assumes loose time synchronization among WAN nodes and does not offer nonrepudiation. If the former is not possible or the latter is desired, an alternative approach is necessary. To this end, they have constructed a secure route discovery protocol (SRDP) that allows the source to securely discover an authenticated route to the destination using either aggregated message authentication codes (MACs) or multisignatures. Several concrete techniques are presented, and their efficiency and security are compared and evaluated.

After a brief discussion on the cryptographic approach the next subsection comprises a detailed study of the trust-based approach of routing.

3.3.2 Trust-Based Approach

In the trust-based approach the source node evaluates the trust level of each intermediate node before establishing any path to the destination node. During broadcast of RREQ, the source node also broadcasts the value of trust level. The nodes with less trust level value are not allowed to take part in route creation. Here, some of the published works on the trust-based approach of secure routing are described.

Argyroudis and O'Mahony [23] have used security attributes as parameters for ad-hoc route discovery in Security-Aware Ad-Hoc Routing (SAR). SAR enables the use of security as a negotiable metric to improve the relevance of the routes discovered by ad-hoc routing protocols. In SAR, the security metric is embedded in the RREQ packet itself. The forwarding behavior of the protocol with respect to RREQs is also changed. Intermediate nodes receive a RREQ packet with a particular security metric or trust level. SAR ensures that this node can only process the packet or forward it if the node itself can provide the required security or has the required authorization or trust level. If the node cannot provide the required security, the RREQ is dropped. If an end-to-end path with the required security attributes can be found, a suitably modified RREP is sent from an intermediate node or the eventual destination. SAR can be implemented based on any on-demand ad-hoc routing protocol with suitable modification.

Buchegger and Boudec propose the idea of CONFIDANT [26]. The basic idea is to make noncooperative nodes unattractive for other nodes to communicate with. A node chooses a route based on trust relationships built up from the experienced, observed, or reported routing and forwarding behavior of other nodes. Each node observes the behavior of all nodes located within the radio range. When a node discovers a misbehaving node, it informs all other nodes in the network by flooding an alarm message. As a result, all nodes in the network can avoid the misbehaving node when choosing a route. This method, however, is prone to false alarm generation by the attacker that makes an innocent node look like a malicious node.

In trusted aodv (TAODV) route selection is based on quantitative route trust and node trust values [26]. Authors have defined route trust from a source node to a destination node as the difference between the number of packets sent from the source node and the number of related packets received by the destination node. Route trust is thus 0 for a perfect route, and trustworthiness decreases for increasing route trust values. TAODV has less per packet overhead. TAODV does not confirm secure routing, as its packet information is not secure itself.

Two different approaches for secure data transmission in WANs have already been discussed. Now, the third approach remains. In the next subsection, message authentication is described.

3.3.3 Message Authentication

This approach has similarities with the cryptographic approach. In both cases, the authentication key or signature is used for verification. The difference in these two approaches is that in the cryptographic approach each packet should carry the authentication key. In the message authentication approach only control packets are supposed to carry the security key. While establishing a route with the destination node, the source node broadcasts RREQ with the security key. Only the target node can send RREP to the source node. Unauthenticated nodes cannot send back any RREP to the source node. In this way, the source node can generate a secure route to its destination node. Here, some earlier works on this approach are discussed.

Papadimitratos and Haas proposed authenticated routing for ad-hoc networks (ARAN) to detect and protect against malicious actions by third parties and peers in an ad-hoc environment. ARAN introduces authentication, message integrity, and nonrepudiation to an ad-hoc environment [21]. ARAN is composed of two distinct stages. The first stage is simple and requires little extra work from peers beyond traditional ad-hoc protocols. Nodes that perform the optional second stage increase the security of their route, but incur additional cost for their ad-hoc peers who may not comply (e.g., if they are low on battery resources). ARAN requires that nodes keep one routing table entry per source-destination pair that is currently active. This is certainly more costly than per destination entries in nonsecure ad-hoc routing protocols.

In secure efficient distance vector routing for mobile wireless ad-hoc networks, Hu et al. [25] stated that the secure efficient ad-hoc distance vector routing protocol (SEAD) is robust against multiple uncoordinated attackers creating an incorrect routing state in any other node, in spite of active attackers or compromised nodes in the network. SEAD is used for the nodes with limited CPU processing capability. This protocol provides a guard against denial-of-service (DoS) attacks in which an attacker attempts to cause other nodes to consume excess network bandwidth or processing time.

Patwardhan et al. [22] have proposed the secure routing protocol (SRP). It is a routing algorithm where they used DSR to design SRP as an extension header that is attached to RREQ and RREP packets. SRP does not attempt to secure RERR packets but instead delegates the route maintenance function to the secure route maintenance portion of the Secure Message Transmission protocol. SRP uses a sequence number in the REQUEST to ensure freshness, but this sequence number can only be checked at the target. SRP requires a security association only between communicating nodes and uses this security association just to authenticate RREQ and RREP through the use of message authentication codes. At the target, SRP can detect modification of the RREQ, and at the source, SRP can detect modification of the RREP. Because SRP requires a security association only between communicating nodes, it uses extremely lightweight mechanisms to prevent other attacks. For example, to limit flooding, nodes record the rate at which each neighbor forwards RREQ packets and gives priority to REQUEST packets sent through neighbors that less frequently forward REQUEST packets. SRP authenticates RREPs from intermediate nodes using shared group keys or digital signatures. When a node with a cached route shares a group key with (or can generate a digital signature verifiable by) the initiator of the REQUEST, it can use that group key to authenticate the REPLYS. The authenticator, which is either a message authentication code, computed using the group key, or a signature, is called the intermediate node reply token. The signature or MAC is computed over the cache REPLY.

Kasiviswanath et al. have proposed Secured AODV (SAODV) [20]. It is an extension of the AODV routing protocol that can be used to protect the route discovery mechanism providing security features like integrity, authentication, and nonrepudiation. SAODV assumes that each ad-hoc node has a signature key pair from a suitable asymmetric cryptosystem. Further, each ad-hoc node is capable of securely verifying the association between the address of a given ad-hoc node and the public key of that node. Achieving this is the job of the key management scheme. Two mechanisms are used to secure the AODV messages: digital signatures to authenticate the nonmutable fields of the messages and hash chains to secure the hop count information (the only mutable information in the messages).

Hu et al. implement two different approaches of secure binding between IPv6 addresses. These are independent of any pre-existing trusted security service. The first method uses signed evidence produced by the originator of the message while the other uses signature verification by the destination, without any form of delegation of trust. The SecAODV implementation follows Tuominen's design, which uses two kernel modules, ip6_queue and ip6_nf_aodv, and a user space daemon AODV. The AODV daemon then generates a 1024-bit Rivest-Shamir-Adleman (RSA) key pair. Using the public key of this pair, the securely bound global and site-local IPv6 addresses are generated. The AODV protocol is comprised of two basic mechanisms, route discovery and maintenance of local connectivity. The SecAODV protocol adds security features to the basic AODV mechanisms, but is otherwise identical. A source node that requests communication with another member of the WAN, referred to as a destination D, initiates the process by constructing and broadcasting a signed route request message RREQ. The format of the SecAODV RREQ message differs from the one proposed by Blum and Eskandarian [44]. It additionally contains the RSA public key of the source node S and is digitally signed to ensure authenticity and integrity of the message. Upon receiving a RREQ message, each node authenticates the source S, by verifying the message integrity and by verifying the signature against the provided public key. Upon successful verification, the node updates its routing table with S's address and the forwarding node's address. If the message is not addressed to it, it rebroadcasts the RREQ.

Papadimitratos and Haas [21] have proposed in "Secure Link State Routing for Mobile Ad hoc Networks" that the Secure Link State Protocol (SLSP) for mobile ad-hoc networks is responsible for securing the discovery and distribution of link state information. The scope of SLSP may range from a secure neighborhood discovery to a network-wide secure link state protocol. SLSP nodes disseminate their link state updates and maintain topological information for the subset of network nodes within *R* hops, which is termed their *zone*. Nevertheless, SLSP is a self-contained link state discovery protocol, even though it draws from, and naturally fits within, the concept of hybrid routing. To counter adversaries, SLSP protects link state update (LSU) packets from malicious alteration, as they propagate across the network. It disallows advertisements of nonexistent, fabricated links, stops nodes from masquerading their peers, strengthens the robustness of neighbor discovery, and thwarts deliberate floods of control traffic that exhaust network and node resources. To operate efficiently in the absence of a central key management, SLSP provides for each node to distribute its public key to nodes within its zone. Nodes periodically broadcast their certified key, so that the receiving nodes validate their subsequent link state updates. As the network topology changes, nodes learn the keys of nodes that move into their zone, thus keeping track of a relatively limited number of keys at every instance. Each node is equipped with a public/private key pair, namely E_v and D_v. SLSP defines a secure neighbor discovery that binds each node *V* to its medium access control address and its *IP* address, and allows all other nodes within transmission range to identify a node unambiguously, given that they already have E_V. Nodes advertise the state of their incident links by broadcasting periodically signed LSUs. SLSP restricts the propagation of the *LSU* packets within the zone of their origin node. Receiving nodes validate the updates, suppress duplicates, and relay previously unseen updates that have not already propagated *R* hops. Link state information acquired from validated *LSU* packets is accepted only if both nodes incident on each link advertise the same state of the link.

Hu et al. have proposed the Secure On-Demand Routing Protocol for Ad-Hoc Networks (ARIADNE) using the TESLA [25] broadcast authentication protocol for authenticating routing messages, since TESLA is efficient and adds only a single message authentication code (MAC) to a message for broadcast authentication. Adding a MAC (computed with a shared key) to a message can provide secure authentication in point-to-point communication; for broadcast communication, however, multiple receivers need to know the MAC key for verification, which would also allow any receiver to forge packets and impersonate the sender. Secure broadcast authentication thus requires an asymmetric primitive, such that the sender can generate valid authentication information, but the receivers can only verify the authentication information. TESLA differs from traditional asymmetric protocols such as RSA in that TESLA achieves this asymmetry from clock synchronization and delayed key disclosure, rather than from computationally expensive one-way trapdoor functions.

Ming-Yang [19] has proposed the wormhole-avoidance routing protocol (WARP), the purpose of which is to defend against wormhole attacks. WARP does not allow any nodes except the destination node to reply to the RREQ with an RREP to the source node. If an intermediate node replies to the RREQ with an RREP, any nodes on the path cannot accumulate the anomaly value of their next neighboring nodes along the route. WARP makes the neighbors of a wormhole node identify that they have abnormal route acquisitions. The neighboring nodes gradually isolate the malicious node. In WARP each node maintains a routing table consisting of various information, like hop count, IP address, destination, and expiration time. Each time the position of the node changes the routing table should be updated. Transmission of huge data and maintenance of such a routing table create overhead in this protocol.

Ammayappan et al. [18] have proposed a protocol in “A New Secure Route Discovery Protocol for MANETs to Prevent Hidden Channel Attacks” to secure a route discovery process, by implementing security mechanisms to protect hidden channels and prevent hidden channel attacks. The mechanism consists of maintaining a monitor table for each of the nodes. Correctness of the route reply (RREP) packet propagation is monitored by verifying the routing information sent by the predecessor node with the stored information in its monitor table. Verification is done with respect to the contents of the monitor table. Information on the identified malicious node is propagated by broadcasting an alarm message. The alarm acceptance message must be sent by upstream nodes of the discovered path, excluding the alarm sender. Based on the successful verification of both alarm and alarm acceptance messages, the malicious node is removed from the authentic neighborhood of its neighbors. In case the RREP packet reaches the source node before the arrival of the alarm and alarm acceptance messages, the source immediately holds back or terminates the path based on the validity of the received alarm and alarm acceptance messages. If no such alarm messages are reported, then it indicates that the path from the source to the target is plausible and secure with respect to various types of wormhole and hidden channel attacks. This approach incurs communication overhead.

3.4 Conclusion

In this chapter, a detailed discussion on routing algorithms for WANs has been given. Some distinguishable features of WAN prevent it from using traditional wired routing protocols. In WAN, there can be three kinds of routing protocols: proactive, reactive, and hybrid. All three approaches have been discussed in detail with some present works. Next, the focus was on security issues in WAN routing protocols. Lack of centralized authority and an infrastructure-less environment make WAN vulnerable to many attacks. So, routing in WAN needs more security. Some existing secure routing protocols were also discussed in this chapter.

References

1. G. Siva Kumar, M. Kaliappan, and L. Jerart Julus, Enhancing the Performance of MANET Using EESCP, in *Proceedings of IEEE International Conference on Pattern Recognition, Informatics and Medical Engineering*, Honolulu, HI, 2012, pp. 225–230.
2. Y. Fu, J. He, L. Luan, G. Li, and W. Rong, A Key Management Scheme Combined with Intrusion Detection for Mobile Ad hoc Networks, in *Proceedings of KES-AMSTA'08*, Inha University, Korea, 2008, pp. 584–593.
3. F.H. Wai, Y.N. Aye, and N.H. James, Intrusion Detection in Wireless Ad-Hoc Networks, in *Proceedings of MobiCom'00*, Boston, MA, 2000, pp. 275–283.
4. Y. Yao, L. Zhe, and L. Jun, Research on the Security Scheme of Clustering in Mobile Ad hoc Networks, in *Proceedings of ITCS'09*, Kiev, Ukraine, 2009, pp. 518–521.
5. J. Parker, J. Undercoffer, J. Pinkston, and A. Joshi, On Intrusion Detection and Response for Mobile Ad-Hoc Networks, in *Proceedings of IPCCC'04*, Concordia, Chicago, 2004, pp. 747–752.
6. M. Shao, J. Lin, and Y. Lee, Cluster-Based Cooperative Back Propagation Network Approach for Intrusion Detection in MANET, in *Proceedings of CIT'10*, Beijing, China, 2010, pp. 1627–1632.
7. N. Marching and R. Datta, Collaborative Technique for Intrusion Detection in Mobile Ad hoc Network, *Ad hoc Networks*, 6(4), 508–523, 2000.
8. P.K. Suri and Kavita Taneja, Exploring Selfish Trends of Malicious Devices in MANET, *Journal of Telecommunications*, 2(2), 25–30, 2010.
9. G. Varaprasad, S. Dhanalakshmi, and M. Rajaram, New Security Algorithm for Mobile Adhoc Networks Using Zonal Routing Protocol, *UBICC Journal*, 2009.
10. Viren Mahajan, Maitreya Natu, and Adarshpal Sethi, Analysis of Wormhole Intrusion Attacks in Manets, in *Military Communications Conference*, San Diego, CA, 2008, pp. 1–7.
11. A. Abraham, R. Jain, J. Thomas, and S.Y. Han, D-SCIDS: Distributed Soft Computing Intrusion Detection System, *Journal of Network and Computer Application*, 30, 81–98, 2007.
12. M. Eid, H. Artail, A. Kayssi, and A. Chehab, An Adaptive Intrusion Detection and Defense System Based on Mobile Agents, at Innovations in Information Technologies, Dubai, 2004.
13. C. Siva Ram Murthy and B.S. Manoj, *Ad hoc Wireless Networks, Architectures and Protocols*, Pearson, Old Tappan, NJ, 2005.
14. C.E. Perkins, *Ad hoc Networking*, Addison Wesley, Reading, MA, 2008.
15. Yogesh Chaba and Naresh Kumar Medishetti, Routing Protocols in Mobile Ad hoc Networks—A Simulation Study Final, *JCS*, 1(1), 83–88, 2005.
16. Aftab Ahmad, *Wireless and Mobile Data Networks*, Wiley Interscience, Hoboken, NJ, 2005.

17. Jihye Kim and Gene Tsudik, SRDP: Secure Route Discovery for Dynamic Source Routing in MANETs, *Ad hoc Networks*, 7(6), 1097–1109, 2009.
18. Kavitha Ammayappan, Vinjamuri Narsimha Sastry, and Atul Neg, A New Secure Route Discovery Protocol for MANETs to Prevent Hidden Channel Attacks, *International Journal of Network Security*, 14(3), 121–141, 2012.
19. S. Ming-Yang, WARP: A Wormhole-Avoidance Routing Protocol by Anomaly Detection in Mobile Adhoc Networks, *Computers and Security*, 29(2), 208–224, 2010.
20. N. Kasiviswanath, S. Madhusudhana Verma, and C. Sreedhar, Performance Analysis of Secure Routing Protocols in Mobile Ad-Hoc Networks, *IJCST*, 3(1), 2012.
21. P. Papadimitratos and Z.J. Haas, Secure Link State Routing for Mobile Ad hoc Networks, in *Proceedings of IEEE Workshop on Security and Assurance in Ad hoc Networks*, IEEE Press, Orlando, FL, 2003, pp. 27–31.
22. A. Patwardhan, J. Parker, M. Lorga, A. Joshi, and T. Karygiannis, Secure Routing and Intrusion Detection in Ad hoc Networks, in *3rd International Conference on Pervasive Computing and Communications*, Kauai Island, HI, 2005, pp. 191–199.
23. Patroklos G. Argyroudis and Donal O'Mahony, Secure Routing for Mobile Ad hoc Networks, *IEEE Communications Surveys and Tutorials*, 7(3), 2–21, 2005.
24. Y. Xue and K. Nahrstedt, Providing Fault-Tolerant Ad-Hoc Routing Service in Adversarial Environments, *Wireless Personal Communications*, 29(3–4), 367–388, 2004.
25. Y.-C. Hu, D.B. Johnson, and A. Perrig, SEAD: Secure Efficient Distance Vector Routing for Mobile Wireless Ad hoc Networks, *Ad hoc Network*, 2, 175–192, 2003.
26. S. Buchegger and J.-Y.L. Boudec, Cooperation of Nodes Fairness in Dynamic Ad-Hoc Networks, in *Proceedings of IEEE/ACM Symposium on Mobile Ad hoc Networking and Computing (MobiHOC)*, Lausanne, Switzerland, 2002, pp. 226–236.
27. Karim El Defrawy and Gene Tsudik, Senior "ALARM: Anonymous Location-Aided Routing in Suspicious MANETs," *IEEE Transactions on Mobile Computing*, 10(90), 1345–1358, 2011.
28. M. Joa-Ng and I-Tai Lu, A Peer-to-Peer Zone-Based Two-Level Link State Routing for Mobile Ad hoc Networks, *IEEE JSAC*, 17(8), 1415–1425, 1999.
29. J.N. Cao, SSR: Segment-by-Segment Routing in Large-Scale Mobile Ad hoc Networks, in *Proceedings of the 3rd IEEE International Conference on Mobile Ad-Hoc and Sensor Systems*, Vancouver, Canada, 2006, pp. 216–225.
30. S. Sivavakeesar and G. Pavlou, Scalable Location Services for Hierarchically Organized Mobile Ad hoc Networks, in *Proceedings of MobiHoc*, Urbana-Champaign, IL, 2005, pp. 217–228.
31. X. Du and D. Wu, Adaptive Cell-Relay Routing Protocol for Mobile Ad hoc Networks, *IEEE Transactions on Vehicular Technology*, 55(1), 278–285, 2006.

32. Yuntao Zhu, Junshan Zhang, and Partel Kautilya, Stochastic Location-Aided Routing for Mobile Ad-Hoc Networks, in *Wireless Communications, Networking and Information Security (WCNIS)*, Beijing, China, 2010, pp. 364–370.
33. Young-Bae Ko and N.H. Vaidya, Location-Aided Routing (LAR) in Mobile Ad hoc Network, *Wireless Networks*, 6(4), 307–321, 2000.
34. S.S Basurra, M. De Vos, J. Padget, T. Lewis, and S. Armour, A Zone-Based Routing Protocol with Parallel Collision Guidance Broadcasting for MANET, in *Communication Technology (ICCT)*, 2010, pp. 1188–1191.
35. M. Abolhasan and T. Wysocki, Dynamic Zone Topology Routing for MANETs, *European Transactions on Telecommunications*, 18(40), 351–368, 2007.
36. S. Corson and J. Macker, *Mobile Ad hoc Networking (MANET): Routing Protocol Performance Issues and Evaluation Considerations*, RFC 2501, 1999, http://www.ietf.org/rfc/rfc2501.txt.
37. B. Bellur and R.G. Ogier, A Reliable, Efficient Topology Broadcast Protocol for Dynamic Networks, in *Proceedings of INFOCOM*, 1999, pp. 178–186.
38. R.G. Ogier, Efficient Routing Protocols for Packet-Radio Networks Based on Tree Sharing, in *Proceedings of the 6th IEEE International Workshop on Mobile Multimedia Comm*unications, 1999, pp. 104–113.
39. Tao Lin and Scott F. Midkiff, *Mobile Ad-hoc Network Routing Protocols: Methodologies and Applications*, 2004, http://hdl.handle.net/10919/11127.
40. R.G. Ogier, F.L. Templin, B. Bellur, and M.G. Lewis, *Topology Broadcast Based on Reverse-Path Forwarding (TBRPF)*, Internet Engineering Task Force (IETF) draft, 2002, http://www.ietf.org/internet-drafts/draft-ietf-manet-tbrpf-06.txt.
41. C. Perkins and P. Bhagwat, Highly Dynamic Destination-Sequenced Distance-Vector Routing (DSDV) for Mobile Computers, in *ACM SIGCOMM'94 Conference on Communications Architectures, Protocols and Applications*, 1994, pp. 234–244.
42. R.S. Kumaran, R.S. Yadav, and K. Singh, Multihop Wireless LAN, *International Journal of Computer Science and Security*, 1(1), 52–69, 2007.
43. S.S. Manvi, M.S. Kakkasageri, S. Paliwal, and R. Patil, ZLERP: Zone and Link Expiry based Routing Protocol for MANETs, *International Journal of Advanced Networking and Applications*, 2(03), 2010, pp. 650–655.
44. J.J. Blum and A. Eskandarian, A Reliable Link Layer Protocol for Robust and Scalable Inter-Vehicle Communications, *IEEE Transactions on Intelligent Transportation Systems*, 8(1), March 2007.

4

Different Types of Attacks for WANs

DEBDUTTA BARMAN ROY
AND RITUPARNA CHAKI

Contents

4.1 Introduction

As discussed in the preceding chapters, a wireless ad-hoc network (WAN) is characterized by its infrastructure-less behavior. Communication between the nodes constituting the WAN is purely on the basis of cooperation and mutual trust. In Chapter 3 it was discussed that WAN works in a completely distributed and decentralized environment. The performance of WAN solely depends on the security and trustworthiness of the nodes in the network. Due to the above-said features, WAN is vulnerable to a wide variety of attacks that target the weakness of WAN [6, 7]. Ad-hoc routing protocols in

WAN, such as dynamic source routing (DSR) and ad-hoc on-demand distance vector (AODV) [15], are prone to some specific kinds of attacks, like blackhole [28], Byzantine [4], and wormhole [2]. Day-to-day demand of wireless networking is increasing in confidential data handling. This is the reason that routing security is one of the hot research areas. As per the International Organization for Standardization/Open Systems Interconnection (ISO/OSI) model, the network conceives of a layer architecture. Each layer performs different functions. On the basis of the distinct functionalities of the layers, the attacking techniques also differ. In this chapter, first a categorization of attacks is done, depending on the attacking techniques and the attacker's location. This is followed by a description of different layer-specific attacks. A more detailed description of some attacks that are considered to be more threatening than others is included in the last part of this chapter. Some preventive approaches are described in Chapters 5 to 7 of this book.

4.2 Security Attacks on WAN

In this chapter we focus on the different attacks in WAN [10, 14], which leads to a brief discussion on security attacks in WAN. This section is concerned with a wide variety of attacks in WAN, depending on the attacking technique and attacker location. The attacker may be present inside the network, which is called an internal attack. When the attacker attacks from outside the network, it is defined as an external attack. If the attacker actively takes part in attacking the network, then it is termed an active attack; otherwise, it is defined as a passive attack. Detailed descriptions of all these attacks are given in this section.

4.2.1 Passive Attacks

In a passive attack, an attacker does not actively participate in decreasing the network performance. The attackers collect the information about the source node, the destination node, and the route established between them. This information is then forwarded to other malicious nodes that in turn effect attacks like denial of service (DoS). The nature of attacks varies greatly from one set of circumstances to another [8, 9].

4.2.2 Active Attacks

The goal of an active attack is to disrupt the packets that are destined for other nodes in the network. Here, the attackers offer an attractive route to the destination node. So, the source node can easily choose that path for packet forwarding. Then the malicious node collects all the packets and destroys them, drops them, or forwards them on a false route. The destination node does not receive the packets sent by the source node.

4.2.3 External Attacks

Here, we are concerned with the location of the attacker in WAN. In the external attack the attacker does not belong to the same domain as the mobile hosts. The attacker attacks the network from outside the domain. External attacks can be prevented by not permitting the external node to access any internal node.

4.2.4 Internal Attacks

In the internal attack the attacker belongs to the same domain as that of other mobile hosts. An internal attack is more vulnerable than an external attack. Prevention of an internal attack is very difficult.

4.3 Layer-Specific Attacks

A network is conceived of layers that are distinguished by their different functionalities. Due to this, each layer faces different types of attack. In this section, layer-wise attacks are described. This chapter is concerned with WAN. Thus, the network, transport, and application layers are considered for discussion.

4.3.1 Network Layer Attacks

In this subsection, a variety of attacks targeting the network layer are identified and briefly discussed.

Routing cache poisoning attack: This can occur only when the nodes in the network are allowed to operate in a promiscuous mode. In the promiscuous mode of operations, whenever a node overhears any new route information, it is supposed to use the information to update its own route cache. This may be useful for efficient route discovery at a later stage. In the routing cache poisoning attack, an intruder in the network broadcast spoofs routing information to poison the route cache of other nodes in that network.

Blackhole attack: The blackhole attack is introduced in [13]. In this attack, a malicious node uses the routing protocol to advertise itself as having the shortest path to the node whose packets it wants to intercept. In a flooding-based protocol such as AODV [11], the attacker listens to requests for routes. When the attacker receives a request for a route to the target node, the attacker creates a reply where an extremely short route is advertised. If the malicious reply reaches the requesting node before the reply from the actual node, a forged route is created. Once the malicious device has been able to insert itself between the communicating nodes, it is able to do anything with the packets passing between them. It can choose to drop the packets to perform a denial-of-service attack, or alternatively, use its place on the route as the first step in a man-in-the-middle attack.

Routing table overflow: In this attack, an attacker attempts to create routes to nonexistent nodes. The goal of this attack is to create a huge amount of false routes that prevent generation of any new route. The table-based proactive routing, where the route is generated before it is needed, is more vulnerable to this kind of attack, as each node in the network maintains separate tables with routing information. Overflow in these tables with spurious, nonexisting routes can create havoc in the network during route discovery.

Sleep deprivation: The sleep deprivation torture is introduced in [28]. Usually, this attack is practical only in ad-hoc networks, where battery life is a critical parameter. Here, the attackers try to consume the battery power of the devices by forwarding unnecessary control packets or route requests. When the power of nodes in the network is exhausted, the nodes go down, implying that they go into the sleep mode and deny service to the network.

Location disclosure: A location disclosure attack can reveal something about the locations of nodes or the structure of the network. The goal of this attack is to leak out information and disclose the security-sensitive location information of nodes that can cause other attacks. It gathers the node location information, such as a route map, and knows which nodes are situated on the target route. When the target route is disclosed, it can be disrupted by other attacks to degrade the performance of the network.

Byzantine attack: A compromised intermediate node works alone, or a setoff compromised intermediate node works in collusion and carries out attacks such as creating routing loops, forwarding packets through nonoptimal paths, or selectively dropping packets, which results in disruption or degradation of the routing services [17].

Rushing attack: In rushing attacks the attacker sends unnecessary route request packets to the nodes in the network so that when an innocent node wants to form a route with the network, it is prevented from doing so by the attacker [19]. The rushing attack can act as an effective denial-of-service attack against all currently proposed on-demand mobile ad-hoc network (MANET) routing protocols, including protocols that were designed to be secure, such as authenticated routing for ad-hoc networks (ARAN) and Ariadne [5].

Eavesdropping: This is one of the easiest means of attack in a wireless sensor network. This involves information gathering in a passive manner [27]. Here the aim of the attacker is to create confusion among nodes in the network by changing the packet's information.

4.3.2 Transport Layer Attacks

After the discussion about network layers, now we focus on transport layer attacks. The transport layer protocols in WAN includes setting up of end-to-end connections, end-to-end reliable delivery of packets, flow control, congestion control, and clearing of end-to-end connections. The transport layer is more vulnerable to the classic SYN flooding attack or session hijacking attacks.

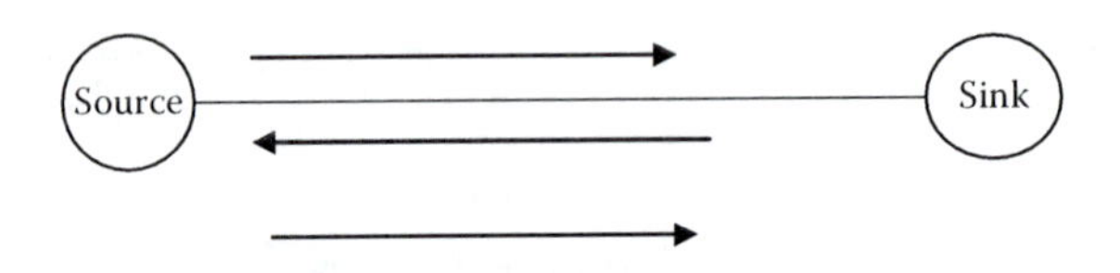

Figure 4.1 TCP handshaking.

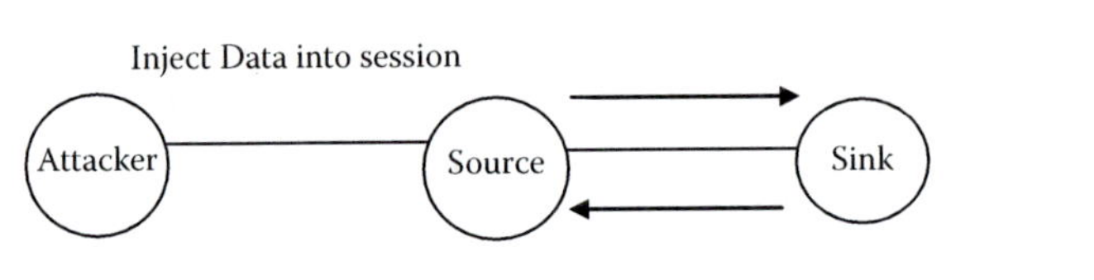

Figure 4.2 Session hijacking.

SYN flooding attack: The SYN flooding attack is a denial-of-service attack. The attacker creates a large number of half-opened transmission control protocol (TCP) connections with a victim node, but never completes the handshake to fully open the connection. In Figure 4.1, it is seen that the nodes are allowed to communicate only when the connection is fully opened. If the connection is half opened, that prevents any further communication.

Session hijacking: Session hijacking takes advantage of the fact that most communications are protected at session setup, but not thereafter (Figure 4.2). In the TCP session hijacking attack, the attacker spoofs the victim's IP address, determines the correct sequence number that is expected by the target, and then performs a DoS attack on the victim. Thus, the attacker impersonates the victim node and continues the session with the target.

4.3.3 *Application Layer Attacks*

The application layer attacks are attractive to attackers because the information they seek ultimately resides within the application, and it is direct for them to make an impact and reach their goals.

Malicious code attacks: Malicious code attacks are caused by viruses, worms, spyware, and Trojan horses. They can attack both operating systems and user applications. These malicious programs usually can spread undesired activity through the network and cause the computer system and network to slow down or even be damaged. In WAN, an attacker can produce attacks similar to those of the mobile system of the ad-hoc network.

4.3.4 Multilayer Attacks

After considering each layer separately, now we discuss those attacks that can occur in any layer, irrespective of their functionality. Examples of multilayer attacks are denial-of-service (DoS), man-in-the-middle, and impersonation attacks.

Denial of service: Denial-of-service (DoS) attacks can be launched from several layers. An attacker can employ signal jamming at the physical layer, which disrupts normal communications. At the link layer, malicious nodes can occupy channels through the capture effect, which takes advantage of the binary exponential scheme in MAC protocols and prevents other nodes from channel access. At the network layer, the routing process can be interrupted through routing control packet modification, selective dropping, table overflow, or poisoning. At the transport and application layers, SYN flooding, session hijacking, and malicious programs can cause DoS attacks.

Impersonation attacks: Impersonation attacks are launched by using another node's identity, such as a MAC or IP address. Impersonation attacks are sometimes the first step for most attacks, and are used to launch further, more sophisticated attacks.

Man-in-the-middle attacks: An attacker sits between the sender and the receiver and sniffs any information being sent between two ends. In some cases, the attacker may impersonate the sender to communicate with the receiver, or impersonate the receiver to reply to the sender.

As mentioned above, this chapter details some of the more vulnerable attacks in WAN. So, the following three sections cover these attacks. In the next section, the blackhole attack is discussed. As the AODV routing is prone to blackhole attacks, we will describe the blackhole attack in WAN.

4.4 Blackhole Attack

In a blackhole attack, a malicious node impersonates a destination node by sending a spoofed route reply packet to a source node that initiates a route discovery. By doing this, the malicious node can deprive the traffic from the source node. A blackhole has two properties. First, the node exploits the ad-hoc routing protocol, such as AODV, to advertise itself as having a valid route to a destination node, even though the route is spurious, with the intention of intercepting packets. Second, the node consumes the intercepted packets. A malicious node always responds positively with a RREP message to every RREQ, even though it does not really have a valid route to the destination node. Since a blackhole does not have to check its routing table, it is the first to respond to the RREQ in most cases. When the data packets routed by the source node reach the blackhole node, it drops the packets rather than forwarding them to the destination node. The goal of intrusion detection is seemingly simple: to detect attacks. However, the task is difficult, and in fact, intrusion detection systems do not detect intrusions at all—they only identify evidence of intrusions, either while they're in progress or after the intrusion [16–18]. Reliable and complete data are required about the target system's activities for accurate intrusion detection. Reliable data collection is a complex issue in itself.

AODV [19] is a reactive routing protocol where the network generates routes at the start of communication. Each node has its own sequence number, and this number increases when connections change. Each node selects the most recent channel information, depending on the largest sequence number. The basic nature of AODV is favorable for a blackhole [20] attack. A malicious node advertises itself as having a valid route to a destination node, even though the route is spurious, with the intention of intercepting packets. In AODV, Dst_Seq is used to determine the freshness of routing information contained in the message from the originating node. When generating a RREP message, a destination node compares its current sequence number and Dst_Seq in the RREQ packet plus one, and then selects the larger one as RREP's Dst_Seq. Upon receiving a number of RREP messages, a source node selects the one with the greatest Dst_Seq in order to construct a route. To succeed in the blackhole attack, the

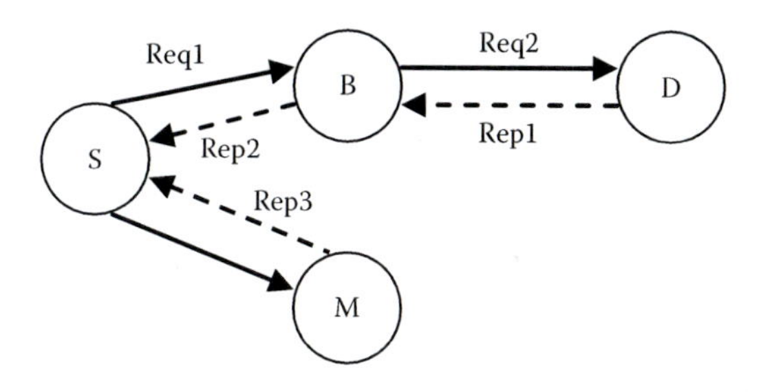

Figure 4.3 Blackhole attack.

Table 4.1 Values of RREQ and RREP

	RREQ		RREP		
	REQ1	REQ2	REP1	REP2	REP3
Intermediate node	S	B	D	B	M
Destination node	D		D		D
Source node	S				
Dst_Seq	11		12		20

attacker must generate its RREP with a Dst_Seq greater than the Dst_Seq of the destination node. It is possible for the attacker to find out the Dst_Seq of the destination node from the RREQ packet. In general, the attacker can set the value of its RREP's Dst_Seq based on the received RREQ's Dst_Seq. However, this Dst_Seq may not be present in the current Dst_Seq of the destination node. Figure 4.3 shows an example of the blackhole attack. The values of RREQ and RREP used in the attack are shown in Table 4.1.

In Table 4.1, the intermediate source node indicates the node that generates or forwards a RREQ or RREP, the destination node indicates the destination node, and the source node indicates the node generated and packet sent. Here, it is assumed that the destination node D has no connections with other nodes. The source node S constructs a route in order to communicate with destination node D. Let destination node D's Dst_Seq that source node S become 11. Hence, source node S sets its RREQ (Req1) and broadcasts as shown in Table 4.1. Upon receiving RREQ (Req1), node B forwards RREQ (Req2) since it is not the destination node. To impersonate the destination node, the attacker M sends spoofed RREP (Rep3), shown in Table 4.1, with the intermediate source node, the destination node the same with D, and increased Dst_Seq (in this case 20) to source node S. At the same time, the destination node D, which received RREQ (Req2), sends RREP (Rep1) with Dst_Seq incremented by one to node S. Although the source node S receives two RREP, based on Dst_Seq the RREP (Rep3) from the attacker M is judged to be the most recent routing information and the route to node M is established. As a result, the track from the source node to the destination node is deprived by node M.

Now that we have briefly discussed the blackhole attack, we turn to another most vulnerable attack. The next section details the wormhole attack, which WAN is very prone to.

4.5 Wormhole Attack

A wormhole attack [12] is a particularly severe attack on MANET routing where two attackers, connected by a high-speed off-channel link, are strategically placed at different ends of a network, as shown in Figure 4.1. These attackers then record the wireless data they overhear, forward the data to each other, and replay the packets at the other end of the network. Replaying the valid network messages at improper places, wormhole attackers can make distant nodes believe that they are immediate neighbors, and force all communications between affected nodes to go through them.

In general, ad-hoc routing protocols fall into two categories: proactive routing protocols that rely on periodic transmission of routing packet updates, and on-demand routing protocols that search for routes only when necessary. A wormhole attack can be dangerous for both proactive and on-demand routing protocols [16, 21–23].

When a proactive routing protocol [24] is used, ad-hoc network nodes send periodic HELLO messages to each other indicating their participation in the network. In Figure 4.2, when node S sends a HELLO message, intruder M1 forwards it to the other end of the network, and node H hears this HELLO message. Since H can hear a HELLO message from S, it assumes itself and node S to be direct neighbors. Thus, if H wants to forward anything to S, it may do so unknowingly through the wormhole link. This effectively allows the wormhole attackers full control of the communication link.

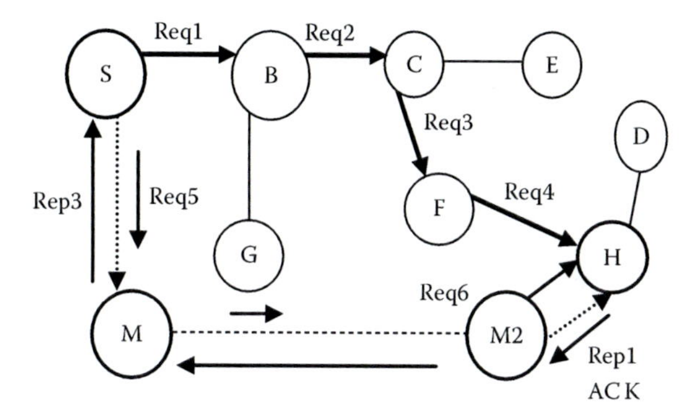

Figure 4.4 MANET with a wormhole attack.

In case of on-demand routing protocols, such as AODV [1, 3], when a node wants to communicate with another node, it floods its neighbors with requests, trying to determine a path to the destination. In Figure 4.4, if S wants to communicate with H, it sends out a request. A wormhole, once again, forwards such a request without change to the other end of the network, possibly directly to node H. A request also travels along the network in a proper way, so H is led to believe it has a possible route to node S through the wormhole attacker nodes. If this route is selected by the route discovery protocol, once again, wormhole attackers get full control of the traffic between S and H. Once the wormhole attackers have control of a link, attackers can drop the packets to be forwarded by their link. They can drop all packets, a random portion of packets, or specifically targeted packets. Attackers can also forward packets out of order or "switch" their link on and off [3].

In Table 4.2, intermediate source node indicates the node that generates or forwards a RREQ or RREP, the destination node indicates the destination node, and the source node indicates the node generated and packet sent.

Table 4.2 Values of RREQ and RREP

	RREQ						RREP		
	REQ1	REQ2	REQ3	REQ4	REQ5	REQ6	REP1	REP2	REP3
Intermediate node	S	B	C	F	M1	M2	H	M2	M1
Destination node	H						H		
Source node	S								

Here, it is assumed that the destination node H has no connections with other nodes. The source node S constructs a route in order to communicate with destination node H. The source node S sets its RREQ_(Req1) and broadcasts as shown in Table 4.1. Upon receiving RREQ_(Req1), node B forwards RREQ_(Req2), since it is not the destination node, to next-hop node C. After receiving RREQ_(Req2) C forwards it to next-hop node F. Node F then sends RREQ_(Req4) to destination node H.

Just now we have discussed those attacks that are prone to network layer only. In the preceding subsection we talked a little bit about the DoS attack. But as it is vulnerable to many layers, we need to gain more knowledge about this type of attack. The next section includes a detailed description of a DoS attack by a selfish node.

4.6 Denial-of-Service Attack

This section deals with the denial-of-service attack (DoS) by a selfish node; this is the most common form of attack that decreases the network performance. In case of DoS attacks, a selfish node is not actually keen to attack the other nodes. However, it does not want to spend its energy, CPU cycles, or available network bandwidth to forward packets not of direct interest to it. It expects other nodes to forward packets on its behalf. The reason behind this is "saving one's own resource" by saving of battery power, CPU cycles, or protecting wireless bandwidth in a certain direction. Therefore, there is a strong motivation for a node to deny packet forwarding to others, while at the same time using the services of other nodes to deliver its own data. According to the attacking technique, the selfish node can be defined in three different ways [25]:

SN1: These nodes participate in the route discovery and route maintenance phases but refuse to forward data packets to save resources.

SN2: These nodes participate in neither the route discovery phase nor the data-forwarding phase. Instead, they use their resources only for transmission of their own packets.

SN3: These nodes behave properly if its energy level lies between full energy level E and certain threshold T1. They behave like

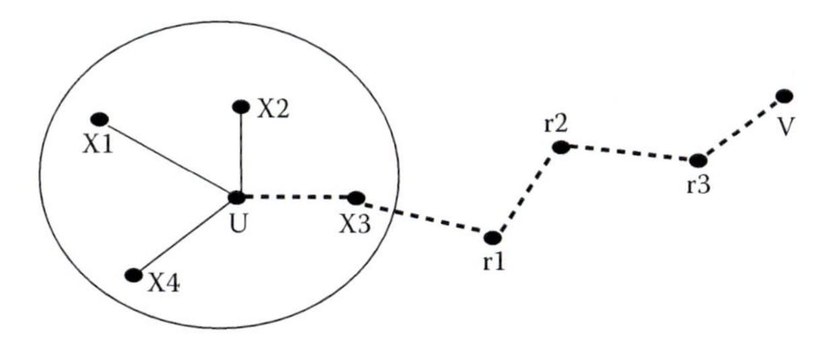

Figure 4.5 Node isolation due to selfish neighbors.

a node of type SN2 if the energy level lies between threshold T1 and another threshold T2, and if energy level falls below T2, they behave like a node of type SN1.

One immediate effect of node misbehaviors and failures in wireless ad-hoc networks is the node isolation problem and network partitioning due to the fact that communications between nodes are completely dependent on routing and forwarding packets [26].

In Figure 4.5, suppose node X3 is a selfish node. When node U initiates a route discovery to another node V, the selfish neighbor X3 may be reluctant to broadcast the route request from u. In this case, X3 behaves like a failed node. It is also possible for X3 to forward control packets; however, the situation could be worse, since u may select X3 as the next hop and send data to it. Consequently, X3 may discard all data to be forwarded via it, and then communications between U and V cannot proceed. When all neighbors of U are selfish, U is unable to establish any communications with other nodes at a distance of more than one hop away. In this case, we say that a node is isolated by its selfish neighbors. Note that selfish nodes can still communicate with other nodes (via their cooperative neighbors), which is different from failed nodes.

4.7 Conclusion

In this chapter we focused on the different types of attacks in wireless ad-hoc networks and classified those based on the attacking techniques and attacker locations. The chapter also discussed attacks on different layers of the networking stack. In order to design an intrusion detection system, researchers should focus not only on attacks but also on the network layers where the attack occurs. As each layer in the network performs different functions, the prevention scheme requires being different for different network layers.

References

1. M. Zapata, Secure Ad hoc On-Demand Distance Vector (SAODV), *Mobile Computing and Communications Review*, 6(3), 106–107, 2002.
2. Y. Hu, A. Perrig, and D. Johnson, Packet Leashes: A Defense against Wormhole Attacks in Wireless Ad hoc Networks, in *Proceedings of IEEE INFOCOM*, 2003, vol. 3, pp. 1976–1986.
3. Y. Hu, D. Johnson, and A. Perrig, SEAD: Secure Efficient Distance Vector Routing in Mobile Wireless Ad-Hoc Networks, in *Proceedings of the 4th IEEE Workshop on Mobile Computing Systems and Applications (WMCSA'02)*, 2002, pp. 3–13.
4. B. Awerbuch, D. Holmer, C. Nita-Rotaru, and H. Rubens, An On-Demand Secure Routing Protocol Resilient to Byzantine Failures, in *Proceedings of the ACM Workshop on Wireless Security*, 2002, pp. 21–30.
5. Y. Hu, A. Perrig, and D. Johnson, Ariadne: A Secure On-Demand Routing for Ad hoc Networks, *Wireless Networks*, 11(1–2), 21–38, 2005.
6. T. Karygiannis and L. Owens, Wireless Network Security—802.11, Bluetooth and Handheld Devices, Special Publication 800-848, National Institute of Standards and Technology, Technology Administration, U.S. Department of Commerce, 2002.
7. C. Kaufman, R. Perlman, and M. Speciner, *Network Security Private Communication in a Public World*, 2nd ed., Prentice Hall, 2002.
8. Djamel Djenouri and Nadjib Badache, Struggling against Selfishness and Black Hole Attacks in MANETs, *Wireless Communications and Mobile Computing*, 8(6), 689–704, 2008.
9. M. Abolhasan and T. Wysocki, Dynamic Zone Topology Routing for MANETs, *European Transactions on Telecommunications*, 18(4), 351–368, 2007.
10. R. Nichols and P. Lekkas, *Wireless Security—Models, Threats, and Solutions*, chap. 7, McGraw-Hill, New York, 2002.
11. Latha Tamilselvan and V. Sankaranarayanan, Prevention of Co-operative Black Hole Attack in MANET, *Journal of Networks*, 3(5), 13–20, 2008.
12. Marianne A. Azer, Sherif M., El-Kassas Abdel Wahab F. Hassan, and Magdy S. El-oudani, *Intrusion Detection for Wormhole Attacks in Ad hoc Networks: A Survey and a Proposed Decentralized Scheme*, ARES, Barcelona, 2008, pp. 636–641.
13. R. Chaki and N. Chaki, IDSX: A Cluster Based Collaborative Intrusion Detection Algorithm for Mobile Ad-Hoc Network, in *Proceedings of the IEEE International Conference on Computer Information Systems and Industrial Management Applications (CISIM)*, Minneapolis, 2007, pp. 179–184.

14. H. Yang, H. Luo, F. Ye, S. Lu, and L. Zhang, Security in Mobile Ad hoc Networks: Challenges and Solutions, *IEEE Wireless Communications*, 11(1), 38–47, 2004.
15. Y. Hu, A. Perrig, and D. Johnson, Rushing Attacks and Defense in Wireless Ad hoc Network Routing Protocols, in *Proceedings of the ACM Workshop on Wireless Security (WiSe)*, 2003, pp. 30–40.
16. Amitabh Mishra, K. Nadkarni, and Animesh Patcha, Intrusion Detection in Wireless Ad hoc Networks, *IEEE Wireless Communication*, 11(1), 48–60, 2004.
17. Yi Li and June Wei, Guidelines on Selecting Intrusion Detection Methods in MANET, in *Proceedings of ISECON, EDSIG*, 2004, vol. 21, pp. 1–17.
18. Yi-an Huang and Wenke Lee, A Cooperative Intrusion Detection System for Ad hoc Network, in *Proceedings of the ACM Workshop on Security of Ad hoc and Sensor Networks (SASN'03)*, 2003, pp. 135–147.
19. C.E. Perkins, E.M.B. Royer, and S.R. Das, Ad hoc On-Demand Distance Vector (AODV), 2000, http://www.ietf.org/internet drafts/draft-ietf-manet-aodv-05.txt.
20. Satoshi Kurosawa, Hidehisa Nakayama, Nei Katol, Abbas Jamalipour, and Yoshiaki Nemoto, *Dynamic Learning Method Detecting Black Hole Attack on AODV-Based Mobile Ad hoc Networks*, Sendai, Miyagi, Japan (received December 19, 2005; revised and accepted January 27 and March 3, 2006).
21. Y.-C. Hu, A. Perrig, and D.B. Johnson, Wormhole Attacks in Wireless Networks, *IEEE Journal on Selected Areas of Communications*, 24(2), 370–380, 2006.
22. H. Yang, H. Luo, F. Ye, S. Lu, and U. Zhang, Security in Mobile Ad hoc Networks: Challenges and Solutions, *IEEE Wireless Communications*, 11(1), 38–47, 2004.
23. Y.-C. Hu and A. Perrig, A Survey of Secure Wireless Ad hoc Routing, *IEEE Security and Privacy Magazine*, 2(3), 28–39, 2004.
24. T. Clausen, P. Jacquet, A. Laouiti, P. Muhlethaler, A. Qayyum, and L. Viennot, Optimized Link State Routing Protocol, at Proceedings of IEEE INMIC, Pakistan, 2001.
25. A.S. Anand and M. Chawla, Detection of Packet Dropping Attack Using Improved Acknowledgement Based Scheme in MANET, *IJCSI International Journal of Computer Science Issues*, 7(4), 12–17, 2010.
26. T.V.P. Sundararajan and A. Shanmugam, Modeling the Behavior of Selfish Forwarding Nodes to Stimulate Cooperation in MANET, *International Journal of Network Security and Its Applications (IJNSA)*, 2(2), 147–160, 2010.
27. Rituparna Chaki, Intrusion Detection: Ad-hoc Networks to Ambient Intelligence Framework, in *Proceedings of the International Conference on Computer Information Systems and Industrial Management Applications (CISIM)*, Krackow, 2010, pp. 7–12.
28. Debdutta Barman Roy, Rituparna Chaki, and Nabendu Chaki, BHIDS: A New, Cluster Based Algorithm for Black Hole IDS, *Security and Communication Network*, 3(2–3), 278–288, 2010.

5

Honesty and Trust-Based IDS Solutions

NOVARUN DEB, MANALI CHAKRABORTY, AND NABENDU CHAKI

Contents

5.1 Introduction

Intrusions in wireless ad-hoc networks are of a fleeting nature. Securing wired local area networks (LANs) is an easier task, as power no longer remains a constrained resource. This allows any number of highly efficient intrusion prevention techniques to be implemented for securing the network. The very insecure nature of the wireless medium makes it open to a vast array of ever-increasing threats. Since wireless networks tend to be ad-hoc, the very nature of the attacks or threats also tends to be the same. Again, one of the most severe resource constraints of such wireless ad-hoc networks is energy or battery life of the nodes forming the network. This major drawback prevents the network operators from deploying computation-intensive intrusion prevention security protocols, as they would take a toll on the battery life of these wireless nodes. Intrusion prevention systems give way to less computation-intensive and energy-efficient security systems—intrusion detection systems.

Intrusion detection systems (IDSs), unlike intrusion prevention systems (IPSs), are also ad-hoc in nature. This implies that IDSs do not burden the network with unnecessary packet exchanges and other computational overheads unless otherwise required. Simply put, an IDS does not secure every packet that is exchanged on the network. Rather, it waits for an intrusion to occur. The effectiveness of an IDS depends on how quickly it detects intrusions after an attack is launched. The sensitivity of an IDS plays a major role in assuring good quality of service (QoS) to applications running on such wireless ad-hoc networks.

A more complicated and efficient system is an intrusion response system (IRS). The difference between an IDS and an IRS is that an

IDS only detects intrusions within the system and generates alerts to the network administrators. All corrective measures that need to be taken are the result of decisions made by network management. An IRS, on the other hand, not only detects intrusions, but also takes corrective measures as a response to intrusion detection. Such responses may include stopping the application traffic from flowing through the network or rerouting traffic through nonmalicious nodes, bypassing the newly detected intruders.

IRS is beyond the scope of this chapter. This chapter discusses two intrusion detection algorithms, HIDS [21] and TIDS [22]. HIDS is an honesty rate-based algorithm, whereas TIDS is a trust-based algorithm for intrusion detection in wireless ad-hoc networks. The subsequent sections of this chapter discuss the algorithms, how they detect intrusions, a comparative analysis between the two algorithms measuring their performance against the same environment settings, and finally implementing these two algorithms on an energy-efficient routing algorithm, ETSeM [20], and discussing the impact that these two algorithms have on any such wireless ad-hoc routing protocol. All simulation results and graphs shown in subsequent sections have been obtained by simulation using the QualNet simulator.

5.2 State of the Art in Wireless Ad-Hoc Networks

In this section a brief review of the existing intrusion detection systems for different wireless networks is presented. The review includes IDSs for mobile ad-hoc networks (MANETs), as many good solutions have matured in this domain over a relatively long period. IDSX [15] is an anomaly-based heterogeneous intrusion detection scheme that was designed on cluster architecture. The authors claimed that, unlike other anomaly-based intrusion detection schemes, the proposed solution takes care of the high false alarm rates. However, the two-step approach of [15] is expensive in terms of energy and resource consumption. Another novel approach based on neural networks and watermarking techniques was proposed in [16]. Like any artificial neural network, the learning process in this approach takes a toll on the energy efficiency of the algorithm. An intrusion detection scheme based on the secure leader election model was proposed in [17]. The lifetime of normal nodes decreases considerably when the number of malicious nodes in the network increases. This is mainly due to the fact that the proposed solution is not lightweight. Also, selfish nodes become leaders in due course of time, as they do not exhaust energy in running the IDS service. Thus, the number of false negatives also increases. A consensus-based intrusion detection scheme was proposed in SCAN [18]. The solution provided in this scheme suffers from high false negatives when the network is static and from high false positives when the network nodes have high mobility. Also, the communication overhead increases with node mobility and the number of malicious nodes. However, as opposed to SCAN, communication overhead was much reduced by BHIDS [19]. This approach is not lightweight, as every node has to maintain a table, and the table updating process is prone to energy consumption.

A survey on trust-based models used for intrusion detection is further provided in this section. Various models have been proposed for sharing resources in a peer-to-peer (P2P) environment. Quite often, these models fail to consider the trust of peers prior to resource sharing. A personalized trust model with reputation and risk evaluation for P2P resource sharing (PET) [1] is one of the highly cited trust models where a peer always trusts itself. Trust in a peer increases slowly but decreases rapidly. In [1], trust is evaluated quantitatively as the combination of two components: reputation and risk. Reputation is a long-term assessment of the behavior of the peer in the past. Risk, on the other hand, is a short-term assessment of the peer's most recent behavior. Trust comprises two reputation components: recommendation and direct interaction. Recommendations dominate trust evaluation when there has been no direct interaction in the past. A weighted evaluation of these components is used in evaluating reputation. Direct interaction information is also used for evaluating the risk component of trust. PET classifies peers based on the QoS provided by them. Four major categories of QoS are good, no response, low-grade, and Byzantine behavior. Nodes are rewarded positively for good behavior only. Nodes are negatively rewarded for the other three categories. The magnitude of negativity decreases from low grade through no response and Byzantine behavior. Risk is evaluated as the amount of negative score earned due to bad services by the peer in a specific time interval.

In [2], Cho et al. have proposed trust management for MANETs using trust chain optimization. Trust is evaluated based on four components: residue energy level and cooperation (QoS trust) and honesty and closeness (social trust). The trust value of a node i is evaluated by a node j as the weighted sum of these four components. Residue energy level and honesty trust component values are binary; the cooperation trust component is a probabilistic value based on the node's behavior in the last update interval; and the closeness component is an integer representing the number of one-hop neighbors of a node. Every node evaluates trust of its one-hop neighbors by observing its behavior to packet forwarding. Trust evaluation is broadcast throughout the network in the form of status exchange messages.

Xiong and Liu proposed a new trust model in PeerTrust [3]. PeerTrust computes the trust of peers in a network as a function of three components. First, a node N becomes trustworthy when other peers who have interacted with N find it to behave normally. Second is the context of satisfaction. It defines the total number of interactions that a node has performed with its peers. Finally, the balance factor of trust is used to reduce the effects of incorrect satisfaction information coming from malicious nodes. A trust metric $T(u)$ for node u is computed as the total satisfaction earned by u and multiplied by the balance factor of each peer and averaged over the total number of interactions that u has participated in. However, PeerTrust fails to capture the most recent malicious behavior of highly reputed nodes. This is taken care of by specifying a sliding window on the timescale. PeerTrust uses the P-Grid algorithm for distribution and aggregation of trust data across a P2P network. A key value is assigned to each peer based on its ID. Each node stores and maintains trust data about one or more peers in the network. As peers can behave maliciously, any intentional false trust data about a peer gets replicated in the local databases of more than one peer. This redundancy has its overhead. Such malicious behavior could be avoided by following a voting by consensus algorithm.

Wang et al. proposed a trust model based on the concept of Honesty. C-index [4] incorporates the past experience a peer node has had with a collaborator. The more the number of trustworthy recommendations from a peer node, the higher should be the credibility of its recommendations. Also, trust models should consider the diversity of trustworthy collaborations. The larger the number of peer nodes with which a node collaborates, the greater is the reliability of its recommendation. Trust depth (TD) in a community of nodes is measured as the number of pure positive feedbacks (PPFs) a node receives from its peer. It is defined as the difference between the number of satisfactory and unsatisfactory feedbacks from that node. Trust breadth (TB) is the number of peers from which a node receives at least one PPF. Based on TD and TB, the C-index of a node is evaluated. The C-index of a node is used in evaluating its trust. It is defined as the number of peers (Z) in a community of N nodes that have sent at least Z PPFs to the node. The C-index mechanism of trust measurement is much more robust, as it is immune to attacks since any single node sending multiple PPFs to a node does not affect its C-index. However, the method remains vulnerable to synergistic attacks. The C-index mechanism fails when the number of attackers is larger than the current C-index of a node.

In [5], Luo et al. have proposed a trust model based on a fuzzy recommendation for MANETs. Trust is defined by three components: past experience, current knowledge about the entity's behavior, and recommendations from trusted entities. The fuzzy trust model centers around a parameter called the local satisfaction degree, (S_{ij}). S_{ij} is the difference between the number of successful and unsuccessful transactions between two nodes i and j. The fuzzy indirect trust model is the generic trust model that evaluates trust from two component values: direct trust and recommendation trust. Direct trust is evaluated by a node on its neighbor as a result of the interactions between them. Recommendation trust depends on the recommendations provided by a neighbor about a distant node. Recommendation trust is evaluated by a node transitively or by consensus. It is evaluated as the combination of the recommendation from the neighbor and the direct trust that the node has in that neighbor. The neighboring node makes a recommendation about the distant node based on what it receives from its neighbors, transitively. The node has direct trust evaluated for all its neighbors, and each neighbor makes a recommendation about the distant node. Consensus recommendation trust is the union of all these trust recommendations. However, recommendations from a highly trusted node remain questionable (e.g., synergistic effect of selfish nodes). Thus, the trust value of a node is computed globally by combining recommendations from all nodes. RFSTrust uses an adjusted cosine function to find the similarity between nodes *i* and *j*. The higher the degree of similarity, the more consistent is the evaluation of trust between the respective nodes compared to other nodes in the network. Thus, it is not a high range of trust values that makes a node's recommendation credible. Rather, credibility of recommendations increases with similarity in rating opinions.

An energy-efficient multipath routing algorithm, ETSeM, has been chosen as an application domain for implementing intrusion detection systems in routing schemes. A brief survey on different types of multipath routing algorithms in different wireless networks is also provided in this section.

The route discovery process and selection of multiple routes is one of the fundamental issues in multipath routing. A meshed multipath routing M-MPR was proposed in [6] to provide mesh connectivity among the nodes. It also uses selective forwarding of packets among multiple paths. The selection is based on the condition of downstream forwarding nodes, and end-to-end forward error checking (FEC) is used to reduce the overhead of retransmitting the packets based on acknowledgment. Besides being energy efficient, higher-throughput achievement has been claimed in [6] compared to any other node disjoint multipath routing protocol.

Another multipath [7] routing for wireless networks combines the idea of clustering and multipath routing. Clustering is used to speed up the routing by structuring the network nodes hierarchically, and multipath routing is used to provide better end-to-end performance and throughput. The solution in [7] is less prone to interference than conventional multipath routing. It is also quite simple, as each path in the CBMPR just passes through the heads of clusters, resulting in a simple cluster level hop-by-hop routing. A reliable and hybrid multipath routing, RHMR for MANET, was proposed in [8]. It uses proactive-like routing for route discovery and reactive routing for route recovery and maintenance.

LIEMRO [9] is another node disjoint multipath routing based on an event-based sensor network to improve QoS in terms of data reception rate, lifetime, and latency. The primary path from the source node to the sink node consists of the nodes with minimum packet transmission cost at each step. Similarly, the second path is established using the second best nodes at each step. Extra routes are only established if they don't decrease the data reception rate at the sink node.

MHRP [10] is a hybrid multipath routing protocol that was designed to properly exploit the inherent hybrid architecture of wireless mesh networks (WMNs). It uses a proactive routing protocol in mesh routers and reactive routing in mesh clients. The client nodes in a wireless environment are often mobile and have fewer resources. MHRP reduces overhead from client nodes by efficient use of the resourceful router nodes toward route discovery and a security mechanism.

Another multipath routing for WMNs was proposed in MRATP [11]. It uses a traffic prediction model based on a wavelet neural network. The main idea of this paper is to set up one primary and some backup paths between a pair of nodes. The primary path is used to transmit the data, until any node on that path generates a congestion signal. Then the backup paths are used to balance the load in the network. It is claimed that [11] reduces end-to-end delay and balances the load of the whole network efficiently.

In [12], a distributed, load balancing multipath routing algorithm has been proposed for wireless sensor networks. The algorithm has two different protocols: one for load balancing and the other a multipath routing protocol. The multipath routing protocol searches multiple node disjoint paths, and then the load balancing algorithm allocates the traffic over each route optimally. The authors of this paper claimed to achieve higher node energy efficiency and lower average delay and control overhead than the other energy-aware routing algorithms.

Another energy-aware multipath routing algorithm for wireless multimedia sensor networks (WMSNs) is proposed in [13]. In this proposed protocol, each node first finds its neighboring nodes, and then builds several partial disjoint paths from the source to the sink node. If any node in the primary path fails for less remaining energy, then the previous hop node of the failed node will find another partially disjoint path to transmit the data. The failed nodes are put into passive states, so that they cannot further interfere in the route selection process. Initially, this protocol builds fewer paths from the source to the sink, and also has lower routing overhead. Thus, it performs better than maximally radio disjoint multipath routing [14] for WMSNs, in which one link failure leads to an alternative route rediscovery process, which increases the routing cost and wastes the energy.

5.3 Description of the Processes

This chapter discusses two of the protocols mentioned in the previous section: HIDS and TIDS. Both these intrusion detection algorithms are similar in the sense that they rank network nodes based on their performance and distinguish between malicious and nonmalicious nodes solely on the basis of this rank. The rank evaluation and updating process is, however, completely different for both algorithms. The working principles of HIDS and TIDS are explained in detail in the following sections.

5.3.1 Honesty Rate-Based Collaborative Intrusion Detection System (HIDS)

HIDS is a collaborative intrusion detection system that has been proposed for mobile ad-hoc networks. Some arbitrarily selected nodes monitor the behavior of peer nodes in the network. Each node in a MANET is attributed with an honesty rate index, called *h-rate*. All nodes join the network with an initial value of 1 for the *h-rate*. The *h-rate* of a node dynamically increases or decreases depending on its behavior. A node is rewarded when it forwards packets for other nodes. In contrast, a node is penalized when it does some malicious act like dropping packets, etc. The *h-rate* for a node N_X is recomputed based on its current *h-rate*, and the rewards or penalty points that it has accrued. This performance of nodes is monitored by its neighboring peers. As a node is selected randomly to monitor other nodes, the task of the intruder to attack the monitor nodes becomes even more difficult. Before describing the rule-based system, it is necessary to define the terminologies used in HIDS in Table 5.1.

Table 5.1 Terminologies Used in HIDS

TERM	INTERPRETATION
H-RATE	Honesty rate for a node in the network
$T_{H\text{-}RATE}$	Threshold honesty rate, below which a node cannot be selected as a monitor node
T_{GEN}	Threshold ratio of new packets that a node is allowed to generate as normal behavior
T_{DROP}	Threshold packet drop ratio that is the maximum permissible drop rate as normal behavior of a node
N_X	Unique identification of each node
PT_Y	Number of packets to node N_Y
PF_Y	Number of packets from N_Y

The following assumptions are made in order to design the proposed rule-based intrusion detection system:

1. A node can overhear activity of other nodes in the link layer provided those nodes are in the direct wireless transmission range.
2. Every node has a unique ID in the network.
3. All packets are to be signed by the private key of the sender using asymmetric key cryptography. This would ensure authenticity and nonrepudiation, as assumed in HIDS.
4. The threshold values are precalculated and set for the entire network.

5.3.1.1 Outline of the Proposed h-Rate Method When a node joins the network, it broadcasts a key request message. In response, it is assigned a set of private and public keys. The public key is assigned in collaboration by a group of N nodes. No node, except the owner itself, knows the private key for any node in the network. The assignment of a shared private key is done by a polynomial secret-sharing scheme [3]. Each node is assigned an *h-rate* (honesty rate) to represent its trust value, which is represented by a nibble $h_0h_1h_2h_3$. Each new node is initialized with an *h-rate* of 0001. The honesty rate of a node is dynamically reassessed and modified by an arbitrary monitor node using rules 1 through 5, as mentioned in the next section.

Monitoring of the network is performed in a random order, at a random time interval. All data and control packets are signed by the source of the packet in order to display its ownership. Any packet transmitted will contain the *h-rate* of the sender, so the receiver can decide what to do with the packet. Thus, any packet of the network in this method will be sent with the following additional fields in the header:

Unique id of the sender	4 bit h-rate

If a node N_X has *h-rate* < $T_{H\text{-}RATE}$, then the node can increase it by simply forwarding packets. A threshold value, T_{GEN}, is used to monitor malicious activity, above which a node is not allowed to generate packets. Each node N_X in the network stores a neighborhood data set NS_X in the form of $\{N_Y, PT_Y, PF_Y\}$, where N_Y is a neighbor of N_X, PT_Y is the number of packets forwarded to N_Y, and PF_Y is the number of packets received from N_Y. These data would be utilized by the monitoring nodes to award or penalize the nodes in the neighborhood. The rules for monitoring have been presented in the next section.

5.3.1.2 Monitoring the Node Behavior The success of the proposed *h-rate* schema depends on detecting the malicious activities in the network and on the dynamic revision of the *h-rate* of the participating nodes. The *h-rate* for honest nodes is to be increased, while that for suspected or compromised nodes must be lowered. A node whose *h-rate* is higher than a preset threshold $T_{H\text{-}RATE}$ is randomly selected as a monitor node to assess the honesty of its neighboring node. A node in the monitoring mode is given access to the neighborhood data structure of the nodes it has to monitor. The *revise-hrate()* of a node will be dynamically revised as a function of its existing *h-rate* and the award/penalty received. The reward for the honest node and the penalty for a malicious or suspected node will both be controlled by the common *argument h-factor* to the *revise-hrate()* function. Rules 1 to 5 describe the activities of nodes while monitoring neighborhoods.

Rule 1: A node N_X that behaves honestly and forwards packets for other nodes is to be encouraged by awarding positive credit. The award will be proportional to the number of packets forwarded. *A node N_X will be awarded if it forwards packet from other nodes.*

Reward: An honest node N_X is rewarded with a positive number proportional to the number of packets that N_X forwards to its neighbors. The unit of reward is *r* per packet forwarded:

$$h\_factor = r \times \Sigma PT_Y$$

Rule 2: A compromised node N_X may drop packets from other nodes. This will leave the node to receive a greater number of packets than the number of packets forwarded, i.e., ΣPF_Y will be greater than ΣPT_Y, for all $N_Y \in NS_X$. However, some of the packets that a node receives from its neighbors may also be meant for itself. Therefore, if a node is found to drop packets more than a threshold value, only then may it be suspected and penalized. A node N_X will be suspected of dropping packets if $(\Sigma PF_Y - \Sigma PT_Y)/\Sigma PF_Y \geq T_{DROP}$, for all the neighbors $N_Y \in NS_X$, where T_{DROP} is a preset threshold ratio for maximum permissible packet drops as normal behavior of the nodes.

Penalty: A suspected node will be assigned a negative reward, proportional to the ratio of dropped packets. The higher the packet drop, the greater the negative number the reward value will be.

$$h\_factor = -r \times ((\Sigma PF_Y - \Sigma PT_Y)/\Sigma PF_Y) \times 10$$

Rule 3: A suspected node N_X is identified as a blackhole when it drops 100% of the packets. A blackhole is to be detected and penalized accordingly. A node N_X will be detected as a blackhole if $\Sigma PT_Y = 0$, for all the neighbors $N_Y \in NS_X$, i.e., the ratio $(\Sigma PF_Y - \Sigma PT_Y)/\Sigma PF_Y$ would be 1 for blackhole nodes.

Penalty: A blackhole node will be assigned a constant negative $h\_factor = -10r$. The penalty is consistent with that for packet dropping nodes. This negative reward value is set such that the *h-rate* value of a blackhole node falls below $T_{H\text{-}RATE}$.

Rule 4: A compromised node N_X may intentionally try to flood the network by generating a large number of packets, and thereby consuming the bandwidth. In such cases, a neighboring node of N_X, say N_Y, would receive a much higher number of packets than N_X has forwarded to N_Y. This is a case that is somewhat reverse in logic to rule 1. If a node is found to generate packets at an abnormal rate and more than a preset threshold value T_{GEN}, only then would it be suspected as a malicious node and penalized. *A node N_X will be suspected as generating spurious packets to N_Y if $(NS_Y.PF_X - NS_X.PT_Y)/NS_Y.PF_X \geq T_{GEN}$, where T_{GEN} is a preset threshold ratio for maximum permissible ratio of new packets that a node may generate as normal behavior.*

Penalty: A suspected packet generating node will be assigned a negative reward, proportional to the ratio of packets generated spuriously.

$$h\_factor = -r \times ((NS_Y \cdot PF_X - NS_X \cdot PT_Y)/NS_Y \cdot PF_X) \times 10$$

Rule 5: A compromised node N_X may manipulate its own neighbor set and increase the value stored in the PT_Y field to make false claim of positive rewards. This must be detected and such malicious nodes are to be penalized. The fact can be verified by looking up the neighborhood set NS_Y of node

N_Y. A node N_X will be suspected of making a false claim of packet forwarding to N_Y if ($NS_X \cdot PT_Y > NS_Y \cdot PF_X$) for any of the neighboring node $N_Y \in NS_X$.

Penalty: A suspected node that makes a false claim of packet forwarding may have been doing so for a long time. Hence, it will be assigned a fixed negative reward equal to a blackhole node, i.e.,

$$h\_factor = -10r$$

Thus, rule-based monitoring of the behavior of the MANET nodes is done in such a manner that the *h-rate* of a node does not fluctuate abruptly unless it is definitely detected to be a malicious node. For example, a node suspected to be dropping packets is gradually penalized. The node will keep on losing its *h-rate* as it continues to drop packets over a long period of time. However, if it is detected as a blackhole, the *h-rate* degrades sharply. The mechanism thus has an inbuilt protection against both false alarms and serious malfunctioning caused by the intruder.

5.3.1.3 Algorithm for Intrusion Detection

1. Initialize *h-rate* as 0001 for any new node joining the network.
2. Select a node, say X, randomly as a monitor node.
3. Check if *h-rate*(X) > $T_{H\text{-}RATE}$.
4. If not, then go to step 2 and repeat.
5. Node accesses the neighborhood set of its neighbor, say Y.
6. Node X submits *h-factor* for its neighbor Y.
7. Compute the revised *h-rate* for Y using

$$h\text{-}rate^{t+1} = f(h\text{-}rate^{t}, h\text{-}factor)$$

8. Check if *h-rate* for all neighbors of X revised.
9. If not, go to step 5.
10. A node N_X will be awarded if it forwards packets from other nodes.
11. A node N_X will be suspected as dropping packets if $(\Sigma PF_Y - \Sigma PT_Y)/\Sigma PF_Y \geq T_{DROP}$, for all the neighbors $N_Y \in NS_X$, where T_{DROP} is a preset threshold ratio for maximum permissible packet drops as normal behavior of the nodes.

12. A node N_X will be detected as a blackhole if $\Sigma PT_Y = 0$ for all the neighbors $N_Y \in NS_X$, i.e., the ratio $(\Sigma PF_Y - \Sigma PT_Y)/\Sigma PF_Y$ would be 1 for blackhole nodes.
13. A node N_X will be suspected of generating spurious packets to N_Y if $(NS_Y.PF_X - NS_X.PT_Y)/NS_Y.PF_X \geq T_{GEN}$, where T_{GEN} is a preset threshold ratio for the maximum permissible ratio of new packets that a node may generate as normal behavior.
14. A node N_X will be suspected of making a false claim of packet forwarding to N_Y if $(NS_X.PT_Y > NS_Y.PF_X)$ for any of the neighboring node $N_Y \in NS_X$.

5.3.1.4 Finding Secure Route Using h-Rate The concept of the honesty rate of nodes can be extended to assess the trust level of routes, and thus to find secure routing paths among any pair of source-destination nodes. The proposed augmentation may be applied over both proactive and reactive routing protocols. In case of reactive routing, the default protocol of the system would return a path between the source N_S and target N_D. We propose to check the *h-rate* values of the intermediate nodes. If the current *h-rate* for any node N_X is lower than $T_{H\text{-}RATE}$, then all routes via N_X are to be discarded. The on-demand routing protocol has to find an alternate path that does not pass through N_X. This idea has been elaborately implemented in the TIDS algorithm as described in the next sections.

5.3.2 Trust-Based Intrusion Detection System (TIDS)

Intrusions need to be detected under varying circumstances. TIDS focuses on two such scenarios where intrusion detection becomes essential. Since the proposed algorithm is trust based, intruders are identified on the basis of their trust values. Intrusion detection is essential during route setup. Good quality of service can be ensured only when the most trusted route is set up between the source and the destination. Thus, trust values of nodes have to be considered when route request and route reply packets are being exchanged. The dynamically changing topology of mobile ad-hoc networks causes the routes between nodes to change frequently. In such a scenario, intrusion detection is even more important, as nodes may change their behavior over time. As long as packets are being sent along a particular route, some intermediate nodes may start behaving selfishly or maliciously. In order to detect such intruders, the IDS algorithms are to be evoked at regular intervals. The network should react differently for destination nodes and intermediate nodes. Whenever a destination node is found to be an intruder, the application may be terminated and the destination node blacklisted. If an intermediate node is found to be an intruder, it should be bypassed and the route reestablished. This intermediate malicious node must also be blacklisted.

Before getting into the details of TIDS, let us consider some of the common attacks. The most largely simulated attack in networking journals is the *blackhole attack*, where a node drops all the packets that are sent through it. However, considering the fact that attackers are intelligent enough, a more practical and realistic attack is the *grayhole attack* or *selective forwarding*. A grayhole node behaves as a good node to increase its reputation within the network. Once it becomes highly reputed, it starts dropping packets. Later, it again increases its reputation and prevents itself from being detected. There is also the *denial-of-service (DoS) attack* that can be implemented in more ways than one. One such implementation is spurious packet generation. The motive behind a DoS attack is to consume the resources of the network so that peers are denied service. Detecting spurious packet generation becomes all the more difficult if the DoS agent is a member on the route from the source to the destination of some application. A standalone node that generates spurious packets is easily detected.

TIDS assumes that both grayholes and DoS agents are on the route from the source to the destination of some application. Ensuring that the route is set up through these intruders is the first phase of any such attack. Keeping these attack scenarios in mind, one can assume that intrusion detection needs to be done only for the nodes that lie on the route between a source and a destination of some application. Nodes that are not part of active applications will not be considered in the intrusion detection process as well. This makes TIDS a lightweight process. All nodes in the network need not execute the intrusion detection algorithm redundantly.

5.3.2.1 Intrusion-Free Route Discovery Intrusion detection is mandatory during the route setup process. The solution proposed in this paper is based on the following working principles:

- Every node maintains trust information about its one-hop neighbors.
- Trust is evaluated as the weighted sum of two components: direct valuation and indirect reference.
- Direct valuation is again a function of two factors: reputation and risk. Reputation is the measure of the long-term evaluation of the behavior of a node. Risk is the valuation of the most recent behavior of the node.
- Indirect reference refers to the recommendations from one-hop neighbors of the target node, which are also neighbors of the valuation node.
- A sliding window is defined on the timescale. The intrusion detection algorithm is executed after every time slice.

Figure 5.1 illustrates the working of TIDS during the route setup process. Route request packets are initiated from the source of an application. The source node 1 sends a trust request packet to all its one-hop neighbors: 1a, 1b, 1c, 1d, 1e, and 2. Every node replies with direct valuation of itself and indirect references about one-hop neighbors who are common to the node itself and the source of the trust request packet. For example, the source node 1 receives direct valuation replies from all one-hop neighbors. Again, node 2 returns

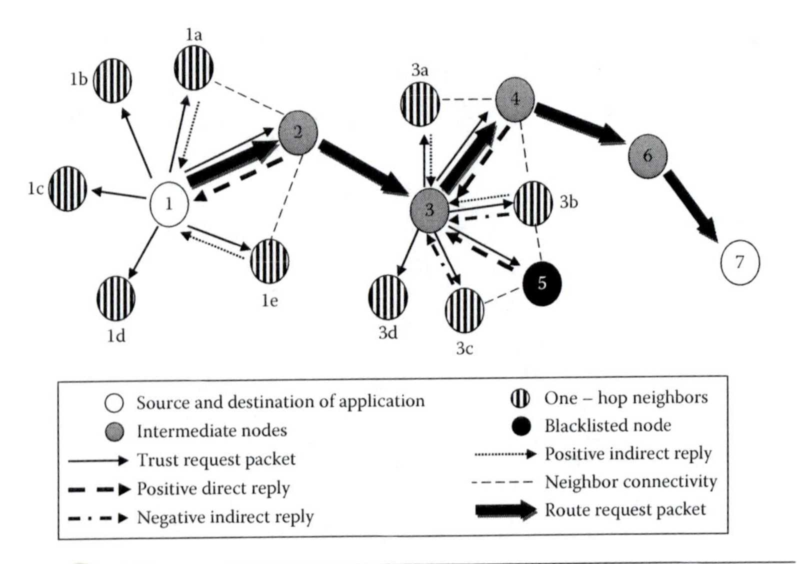

Figure 5.1 Route request mechanism in TIDS. (From N. Deb and N. Chaki, 2012. *Lecture Notes in Computer Science*. With permission.)

indirect references about nodes 1a and 1e, node 1a returns indirect references about nodes 1b and 2, and so on. It is obvious that intruders will speak highly of themselves. Also, attackers may provide incorrect trust information about neighboring nodes in their efforts to establish routes through themselves. Thus, the source of the trust request packets does not believe the responses coming from its one-hop neighbors blindly. Since every node maintains trust information about its one-hop neighbors, the source associates a credibility factor with the replies coming from its neighbors. The most trusted node is evaluated, and the route request packet is forwarded to that node. In the figure, node 2 is found to be the best node on the route from the source to the destination.

This procedure is repeated at every node. Figure 5.1 illustrates another scenario. When the route request packet comes to node 3, the same procedure is repeated as above. The one-hop neighbors of node 3—3a, 3b, 3c, 3d, 4, and 5—reply to the trust request coming from node 3. Node 3 gets positive replies about node 4, but negative replies about node 5. Node 3 associates the credibility of these replies coming from its one-hop neighbors. It evaluates node 4 to be the most trustworthy and node 5 as an intruder. Thus, the route request is forwarded in the direction of node 4. This procedure gets repeated until the route request reaches the destination node.

Figure 5.2 illustrates the procedure when the route request reaches the destination. When the route request reaches node 6, it sends trust request packets to all its neighbors—6a, 6b, 6c, 6d, and 7—including the destination. The destination replies with a route reply and also mentions the number of its one-hop neighbors in the route reply message. All those one-hop neighbors of node 6 that are also neighbors of the destination return their trust information about the destination. Node 6 evaluates the trust of the destination node. If the destination is trustworthy, then the route reply is forwarded to the source; otherwise, a connection abort message is returned. This process prevents malicious intruders from falsely claiming themselves as the destination node of an application.

5.3.2.2 Intrusion Detection and Rerouting Intrusion detection also becomes essential as a part of maintenance. Once a connection is established between the source and the destination, application traffic

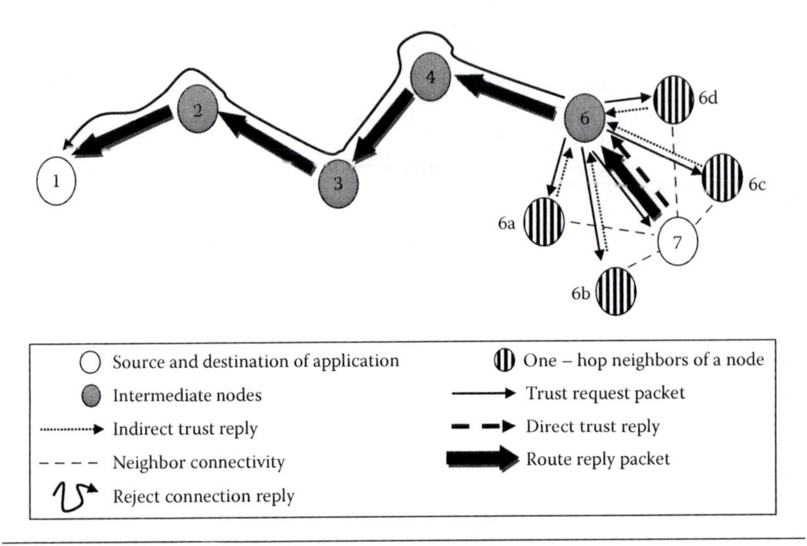

Figure 5.2 Route reply mechanism. (From N. Deb and N. Chaki, 2012. *Lecture Notes in Computer Science*. With permission.)

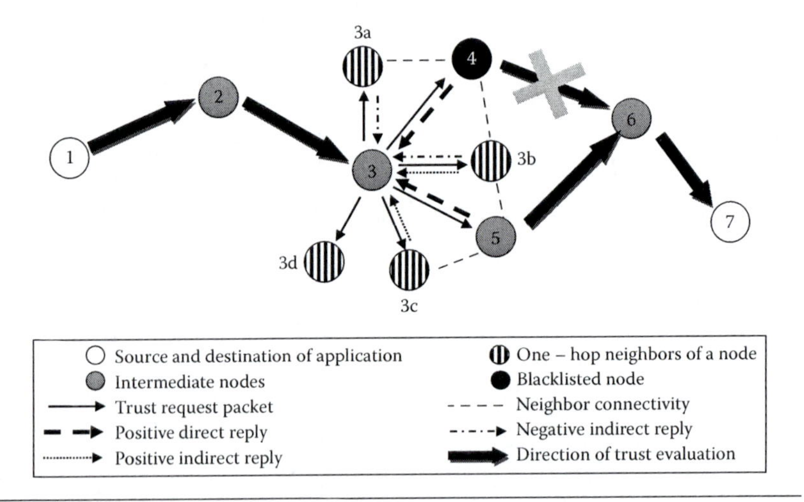

Figure 5.3 Route maintenance mechanism in TIDS. (From N. Deb and N. Chaki, 2012. *Lecture Notes in Computer Science*. With permission.)

starts flowing between the two. An intruder may start behaving maliciously or selfishly at some random time instant. Figure 5.3 illustrates the route maintenance mechanism. Trust evaluation begins at the source. The source evaluates the trust of its one-hop neighbor, which is on the route to the destination. Once trust is evaluated for the one-hop neighbor on the source-destination route, the same procedure is repeated for the next node on the route. This continues until the trust value of the destination node is evaluated.

If, during this process, a node detects its peer on the source-destination route to be behaving selfishly or maliciously, then the rerouting mechanism is initiated. Figure 5.3 best illustrates this mechanism. Suppose the existing route for application traffic is through 1→2→3→4→6→7. During intrusion detection, node 3 finds that node 4 has been behaving in a malignant manner. Node 3 discards the existing route and tries to reroute traffic to the destination, bypassing node 4.

It finds node 5 as trusted and reestablishes connectivity with node 6 via node 5. Thus, the newly established route for application traffic becomes 1◊2◊3◊5◊6◊7. If intrusion detection for maintenance finds that the destination itself has become malicious, then the application is closed.

5.3.2.3 The Trust Model The entire process of routing and intrusion detection is based on the trust evaluated by a node for its one-hop neighbors. The trust model has two main underlying concepts: direct valuation and indirect reference.

Direct valuation is a measure of how the node evaluates the trust of its one-hop neighbors. Every node monitors the packet forwarding behavior of its one-hop neighbors. A benign node should forward all the packets that it receives from its previous hop neighbor. Thus, packet arrival rate (PAR) and packet delivery rate (PDR) play a decisive role in deciding the behavior of a node. For normal node behavior, PAR and PDR tend to be equal. In other words, PAR – PDR tends to zero. Keeping in mind wireless network constraints like mobility and link failure, the normal behavior of a node is classified when the absolute difference (PAR – PDR) lies within a given threshold.

In the selective forwarding attack scenario, a node drops packets occasionally. At other times, it behaves like a normal node. Normal behavior of a node is positively rewarded by increasing that node's trust value. Thus, occasional malicious behavior becomes even more difficult to detect. The proposed IDS addresses this issue using two separate measures: risk and reputation. Risk is a measure of the node's behavior in the last time slice since the last time the intrusion detection algorithm was executed. Reputation is the measure of the long-term behavior of a node. Classifying direct valuation into risk and reputation helps in identifying the most recent behavior of a node in contrast to its long-term behavior on the timescale.

Since direct valuation depends on the PAR and PDR information coming from one-hop neighbors, attackers may easily tamper with this information. Thus, trust of a node is not updated solely on the basis of direct valuation. One also needs to consider the reputation of the target node to all its one-hop neighbors. Thus, indirect references are considered from all those one-hop neighbors that are common to both the evaluation node and the target node. Thus, indirect reference of the evaluation node consists of the reputation information coming from all those one-hop neighbors, which are also neighbors of the target node.

Every node maintains a packet receive (PR) and packet send (PS) counter. After every time slice, these countervalues are sent to the node's one-hop neighbors. The neighbors keep track of the reputation of the node by summing the (PR – PS) differences coming at the end of each time slice. Also, the value of the (PR – PS) counters in the last time slice measures the risk. When a node receives an IDS request packet from an evaluation node, it sends its (PR – PS) countervalues, and reputation and trust information. The (PR – PS) values of the target node are used to evaluate the risk. These values are summed up with the existing reputation data and reputation information of other one-hop neighbors and become the evaluation node's indirect reference information. These three measures are combined to evaluate the reward for the target node's behavior in the last time slice as follows:

$$\text{Reward} = (W1 \times \text{Risk}) + (W2 \times \text{Reputation}) + (W3 \times \text{Indirect reference})$$

The above formula is used to generate negative rewards by assigning negative weights to W1, W2, and W3. Also, these weights are normalized so that $W1 + W2 + W3 = -1$. This formula will be used only when the target node has behaved maliciously in the last time slice, i.e., abs(PAR – PDR) > Threshold. For normal behavior, the reward generated is positive, as follows:

$$\text{Reward} = (\text{PAR} + \text{PDR})/2 * W4$$

W4 is chosen so that the positive reward is not very large. Nodes must not be able to increase their trust values rapidly by behaving normally in some time slices. Once the reward for a node is appropriately calculated, the trust value of the node is updated as follows:

$$\text{Trust}(t) = \text{Trust}(t-1) + \text{Reward}$$

Based on the above formula, the trust of a node may increase gradually or decrease rapidly. Once the trust value of a node is updated, it is checked whether the trust value falls below a certain threshold. If so, then the node is classified as an attacker.

5.3.2.4 Algorithm for Intrusion Detection during Route Setup

1. The source initiates route discovery by generating RREQ packets.
2. Whenever a node receives a RREQ packet, it forwards the packet to the most trusted one-hop neighbor on the route to the destination.
3. The node broadcasts trust request packets to its one-hop neighbors.
4. All neighbors reply with packet forwarding information about themselves and trust information about their one-hop neighbors.
5. The source of the trust request packet evaluates the trust of all its one-hop neighbors.
6. The most trusted neighbor is forwarded the RREQ packet.
7. If a node finds the destination to be its neighbor, it forwards the RREQ packet to the destination.
8. The trust value of the destination is evaluated by the predestination node.
9. The destination responds with an RREP packet. Depending on the trust evaluated of the destination, the predestination node either forwards the RREP packet or returns a "Cancel Application" message toward the source.

5.3.2.5 Algorithm for Intrusion Detection as Part of Maintenance

1. After the time slice expires the source initiates the intrusion detection algorithm.
2. Source sends trust request packet (TRP) to its one-hop neighbors.
3. The one-hop neighbor on the route to the destination returns its packet forwarding information. This is the direct valuation data.
4. Those one-hop neighbors that are not on the route to the destination check if the target node is their one-hop neighbor.
5. If so, they return trust information about the target node to the source of the TRP. This is the indirect reference.

6. The sender of the TRP receives direct and indirect information.
7. Reputation is the trust value of the target node currently available at the source of the TRP. Risk is the packet forwarding information returned by the target node for the last time slice.
8. Indirect recommendations coming from other one-hop neighbors are accumulated, averaged, and combined with results from the previous step.
9. If the target node is found to be an intruder, then a warning message is sent to the source of the TRP that the route is no longer safe.
10. Whenever an intermediate node receives such a message, it reestablishes a new route from itself to the destination.

5.4 Performance Analysis of HIDS and TIDS

Both HIDS and TIDS have been implemented using QualNet. In this simulation, some nodes have been arbitrarily initialized with higher trust values than other nodes. The proposed mechanism successfully sets up routes through the highly trusted nodes. Both grayholes and DoS agents have been implemented as having high initial trust values. This is quite logical, as attackers do try to attain high trust among their peers before launching an attack. The trust value of course changes dynamically during simulation. The data points collected reflect the sensitivity of TIDS compared to HIDS under similar conditions.

5.4.1 Simulator Parameter Settings

In order to compare the performance of HIDS and TIDS in terms of intrusion detection, both solutions have been simulated under the same environment settings. The performance of TIDS is compared with HIDS in detecting different attacks like grayhole and denial of service (DoS). Table 5.2 describes the parameters with which simulation has been done. Trust values of nodes vary from 0 to 16. A trust value of 6 is the threshold value below which a node is detected as an intruder. All normal nodes are initialized with a threshold value of 6. Certain nodes (especially attackers) will have higher trust values. The .config file associated with each QualNet scenario has been suitably modified to assign a trust value of 10 to these highly trusted nodes.

Table 5.2 Simulator Parameter Settings

PARAMETER	VALUE
Terrain area	1500×1500 m^2
Simulation time	200 s
MAC layer protocol	Distributed coordination function (DCF) of IEEE 802.11b standard
Network layer protocol	AODV routing protocol
Traffic model	CBR
Number of CBR applications	10% of total nodes
Highly trusted nodes	Randomly selected
IDS time slice	10 s

5.4.2 Simulation Results

The following data were collected based on the above simulation settings. Four sets of data were collected. The average of these data was taken for comparative analysis of performance between HIDS and TIDS. Data are taken with respect to the number of iterations required for intrusion detection.

Figure 5.4 compares the performance of HIDS and TIDS while detecting denial-of-service attacks. While varying the node density, the percentage of malicious nodes remains fixed. The fewer the number of iterations required to detect all DoS agents, the greater is the sensitivity of the solution. The graph clearly indicates that TIDS is much more sensitive to DoS attacks than HIDS.

The next set of data was taken by varying the node density and implementing the selective forwarding or grayhole attack. Both

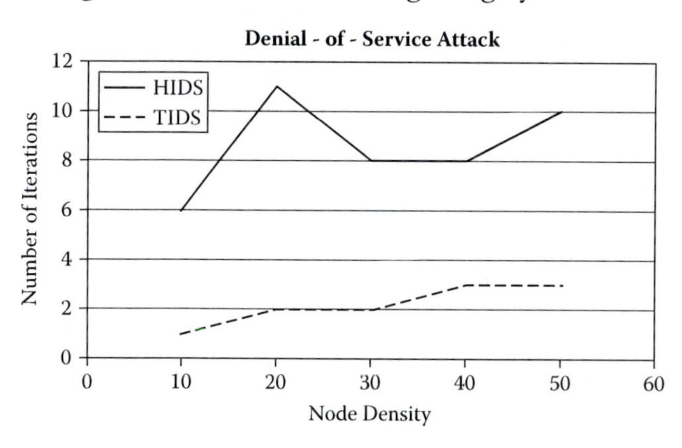

Figure 5.4 Performance for DoS attack with variation in node density. (From N. Deb and N. Chaki, 2012. *Lecture Notes in Computer Science*. With permission.)

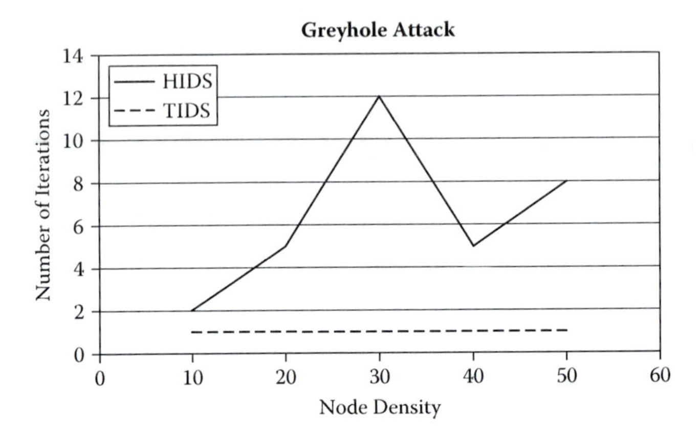

Figure 5.5 Performance for grayhole attack with variation in node density. (From N. Deb and N. Chaki, 2012. *Lecture Notes in Computer Science*. With permission.)

HIDS and TIDS performed reasonably well in terms of false negatives. None of the algorithms generated any false positives. Figure 5.5 shows the results in the form of graphs. Once again, it is seen that TIDS has greater sensitivity to grayhole than HIDS. Also, the curve for TIDS is linear, with the number of iterations required being one. This implies that irrespective of increase in node density, as long as the density of malicious nodes remains the same, TIDS always detects all malicious nodes in the first iteration itself. This is highly impressive.

The next set of data was taken with a fixed number of nodes (= 40) and varying the percentage of malicious nodes. This simulation setting helps in testing the sensitivity of the IDSs when the density of malicious nodes in the network increases.

Figure 5.6 reiterates the fact that TIDS is much more sensitive than HIDS. This is true even when the density of malicious nodes increases. The efficiency of HIDS decreases almost linearly with an increase in malicious node density. However, the performance of TIDS is somewhat constant, as can be seen from the slope of the curve.

Figure 5.7 plots the performances of HIDS and TIDS for the selective forwarding attack or the grayhole attack. The results are again in favor of TIDS. The graphs are very simple to interpret. The sensitivity of HIDS decreases linearly with an increase in grayhole node density, which can be interpreted from the slope of the curve. The sensitivity of TIDS to grayholes remains constant for malicious node

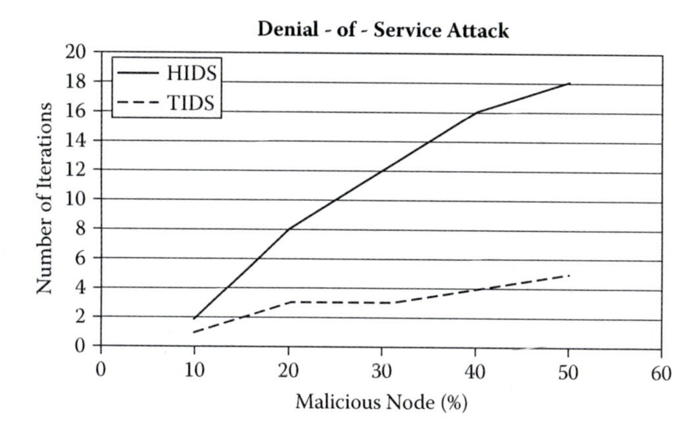

Figure 5.6 Performance for DoS attack with variation in percent of malicious nodes. (From N. Deb and N. Chaki, 2012. *Lecture Notes in Computer Science.* With permission.)

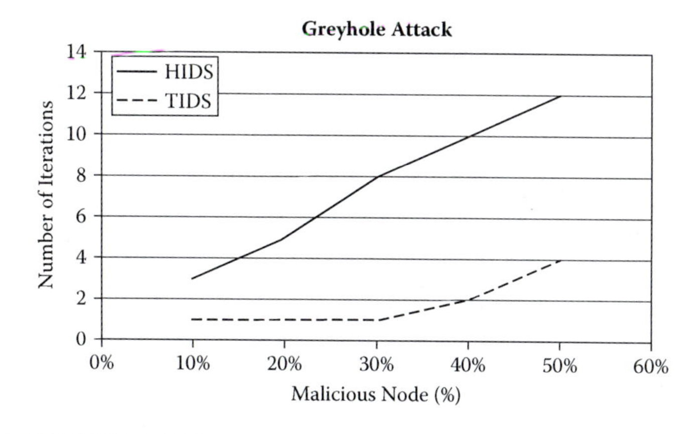

Figure 5.7 Performance for grayhole attack with variation in percent of malicious nodes. (From N. Deb and N. Chaki, 2012. *Lecture Notes in Computer Science.* With permission.)

densities of up to 30%. The sensitivity slightly degrades for higher malicious node densities, but is insignificant compared to that of HIDS.

All results reflected the sensitivity of the trust model in TIDS over the honesty-based scheme proposed in HIDS. These results clearly indicate that TIDS is much more sensitive in detecting malicious behavior than HIDS.

5.5 Implementing IDS in Multipath Routing

The previous sections provide insight into the logic, algorithm, and performance of two standard intrusion detection algorithms: HIDS and TIDS. Merely proposing intrusion detection solutions is not enough. It is always desirable to see how different routing parameters are affected when an intrusion detection process is incorporated into the routing scheme. Since intrusion detection is more desirable in wireless ad-hoc mobile networks with energy constraints, an energy-efficient multipath routing protocol, ETSeM, has been chosen to see the impact of incorporating HIDS and TIDS. Extensive simulation has been done, and the impact of incorporating an IDS has been observed for five different routing parameters: packet delivery ratio, throughput, end-to-end delay, jitter, and average energy consumption of the nodes. The results are very interesting.

5.5.1 Energy-Aware, Trust-Based, Selective Multipath Routing Protocol (ETSeM)

Wireless ad-hoc networks have been developed for on-the-fly reliable data communication and load balancing. Multiple path communication is the basic need behind these two objectives. If these attributes of wireless networks are not utilized properly, one cannot achieve the best out of this network paradigm. Moreover, multipath routing assists in achieving security in routing protocols. Most of the proposed schemes are not able to minimize the overhead of storing extra routes through the lifetime, and the maintenance cost of those routes. ETSeM is one such routing protocol that can manipulate the degree of multipaths according to the energy level, number of paths through a node, and trust value of a node. The working of ETSeM has been described in the following sections. There are mainly two parts: neighbor discovery and route establishment. The working of this algorithm is based on the following principles:

- Every node has to maintain an array HEALTH and two variables, ENERGY and PATH.
- Every node maintains the health information about its one-hop neighbors.
- HEALTH is a linear function of the remaining energy of that node, the number of paths that already exists through the node, and the trust value of the node.
- Every node has to send its energy and path metrics to its one-hop neighbors.

5.5.1.1 Neighbor Discovery At first every node obtains some information about its neighbors. Each node broadcasts a HELLO packet to identify its neighbors within a one-hop distance. On receiving the HELLO packets, each node replies with an ACK packet, containing the remaining energy of that node, the value of a countervariable PATH, which denotes the number of paths already existing through that node and the trust value of that node, given by its neighbors. We assume that every node sent its information reliably. After that, every node calculates the value of the variable HEALTH of its neighbors in terms of their remaining energy and the number of paths already passing through the node.

$$\text{HEALTH} = f(\text{remaining energy, PATH})$$

5.5.1.2 Route Establishment Figure 5.8 shows the route establishment process for this algorithm. To establish a route between a source node and a sink node, the source node sends a ROUTE_REQUEST packet toward the sink node. Each node checks for the healthiest node among its neighbors.

If the value of the variable HEALTH for the healthiest node is greater than 90%, then it forwards the ROUTE_REQUEST packet

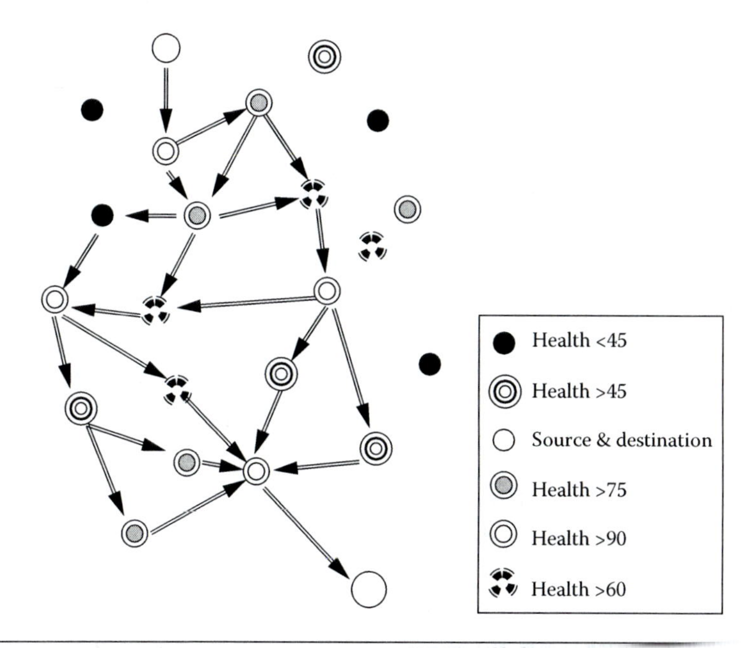

Figure 5.8 Selection of route depending on HEALTH. (From M. Chakraborty and N. Chaki, 2012. *Lecture Notes in Computer Science.* With permission.)

through that node only. If the value of HEALTH is in the range of 75 to 90%, then the sender forwards the ROUTE_REQUEST packet through two nodes, the healthiest one and the second healthiest one. If the value of HEALTH is in the range of 60 to 75%, then the ROUTE_REQUEST packet is forwarded through the first three healthiest nodes in the neighborhood. If the value of HEALTH is in the range of 45 to 60%, then the ROUTE_REQUEST packet is forwarded through the first four healthiest nodes in the neighborhood. Otherwise, that is, when the value of HEALTH is below 45% for every node in the neighborhood, the sender broadcasts the ROUTE_REQUEST packet through all the neighbors.

Each node, after receiving the ROUTE_REQUEST packet, forwards it similarly, and it also keeps the ID of the node from which this packet has been received and inserts its own ID in the packet, to prevent the looping error. A node cannot forward the ROUTE_REQUEST packet to such a node whose ID is already in the packet. When the ROUTE_REQUEST packet reaches the sink node, it replies by transmitting a ROUTE_REPLY packet to the node from which it receives the ROUTE_REQUEST packet. Every node along the path from the sink node to the source node increments the value of the countervariable PATH by one, every time it receives a ROUTE_REPLY packet. Thus, the PATH variable denotes the number of paths passing through a node.

Upon receiving the ROUTE_REPLY packet, the source node confirms a path to the sink node, and uses this path to transmit data. Each node distributes the load equally through all the paths, starting from that node toward the sink node. If some node has only one path toward the sink node, it transmits the data through that path only, whereas if some node has two or more paths toward the sink node, it divides the data equally, and transmits packets through every route. Every node gains some rewards from its neighbors, when it forwards the data packets successfully.

5.5.1.3 Extended ETSeM (E^2TSeM) ETSeM has not been thoroughly tested with different intrusion detection mechanisms. Route selection in E^2TSeM is done based on the remaining energy of the nodes, the number of paths set up through the nodes, and the trust value of the nodes. Incorporating the trust value of the nodes in the route selection process is significant, as can be observed in Figure 5.9. Figures 5.9a

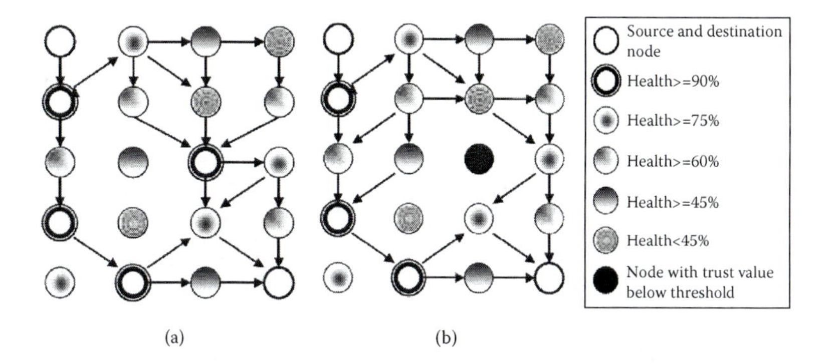

Figure 5.9 (a) Data transmission from Source to Destination using ETSeM; (b) data transmission from Source to destination using E^{2}TSeM. (From M. Chakraborty and N. Chaki, 2012. *Lecture Notes in Computer Science.* With permission.)

and b depict the same set of nodes. The black node in Figure 5.9b has health ≥ 90%, as shown in Figure 5.9a, but has a trust value below the threshold. Hence, the node is malicious, as observed by the underlying security protocol. E^{2}TSeM performs better than ETSeM, as it avoids all such malicious nodes in the route selection process.

This is the extension that is being proposed. No underlying security protocol has been assumed as in the case of ETSeM. Rather, two recent intrusion detection algorithms have been incorporated in the route selection process. First, the honesty-based intrusion detection system (HIDS) has been incorporated into the ETSeM algorithm. This modification is called H-ETSeM. Each node now has an *h-rate* parameter that becomes important in the route selection process.

In the next phase, a trust-based intrusion detection system (TIDS) is also implemented in the ETSeM procedure. This modification is called T-ETSeM. The trust value of the nodes, in this case, plays a role in the route selection process. Both H-ETSeM and T-ETSeM are separately simulated in QualNet.

5.5.2 Simulation Results and Performance Analysis

This section presents a quantitative analysis of ETSeM and how its performance compares with its two modifications, H-ETSeM and T-ETSeM. The simulations have been done extensively. Data have been collected by varying the density of nodes. For each particular node density the experiment has been run five times and the results averaged. The average of these results is then plotted on graphs.

Table 5.3 Simulator Parameter Settings

PARAMETER	VALUE
Terrain area	1500×1500 m^2
Simulation time	100 s
MAC layer protocol	DCF of IEEE 802.11b standard
Traffic model	CBR
Number of CBR applications	10% of the number of nodes
Mobility model	Random waypoint
Initial energy value of normal nodes	5000
Trust value of normal nodes	6–10

ETSeM, H-ETSeM, and T-ETSeM have all been simulated using QualNet. The findings are based on simulation results. The results have been taken by varying the node density from 10 to 50 nodes with a fixed mobility of 30 mps. The simulation scenario and settings are described in Table 5.3.

5.5.2.1 Packet Delivery Ratio The packet delivery ratio is an important and standard metric for any routing algorithm. A constant bit rate (CBR) application is set up for every 10 nodes in the scenario. That is, if the node density of the experiment is 30, we set up three CBR applications. Each CBR application is set up to run for 100 s and to transmit 100 packets. Every time the experiment is run, the CBR server stats are checked. The total number of packets received at the destinations is averaged over the total number of CBR applications.

Figure 5.10 shows the packet delivery ratio for the original ETSeM, H-ETSeM, and T-ETSeM. The packet delivery ratio has a decreasing trend for both ETSeM and H-ETSeM, although ETSeM performs better than H-ETSeM. T-ETSeM, on the other hand, has a more stable and high PDR, ranging between 0.93 and 1.

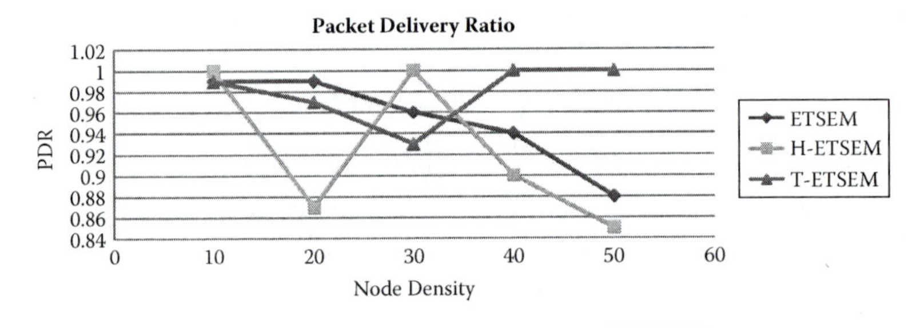

Figure 5.10 Packet delivery ratio vs. node density. (From M. Chakraborty and N. Chaki, 2012. *Lecture Notes in Computer Science*. With permission.)

This demonstrates the fact that ETSeM, when combined with TIDS, performs better multipath routing than ETSeM alone. Incorporating trust values of the nodes in the route selection process proves to be better.

5.5.2.2 Throughput Throughput is measured in kilobits/s. Throughput is obtained from the CBR server stats. The plotted data for node density x is the average of the throughput values over the number of CBR applications, i.e., 10% of x. Figure 5.11 shows the experimental results for throughput. The throughput of ETSeM decreases gradually with an increase in node density. The throughput of H-ETSeM is lower than that of ETSeM alone. However, the throughput for H-ETSeM stabilizes around the 5 Kbps mark. The throughput of T-ETSeM is much better. TIDS is much more efficient than HIDS in terms of overhead and sensitivity. T-ETSeM exhibits much better decision making and efficient path selection, leading to better throughput values.

5.5.2.3 End-to-End Delay The end-to-end delay represents the delay between CBR packets transmitted at the source nodes (CBR clients) and received at the destination nodes (CBR servers). The end-to-end delay values are obtained from the CBR server stats. The delay values for individual CBR servers are summed up and averaged over the number of CBR applications.

Figure 5.12 shows the graph for average end-to-end delay with varying node density. Delay values for ETSeM are somewhat stable over changing node density. The delay values for H-ETSeM have an increasing trend with increasing node density. This is obvious, as greater node density means greater packet overhead for *h-rate*

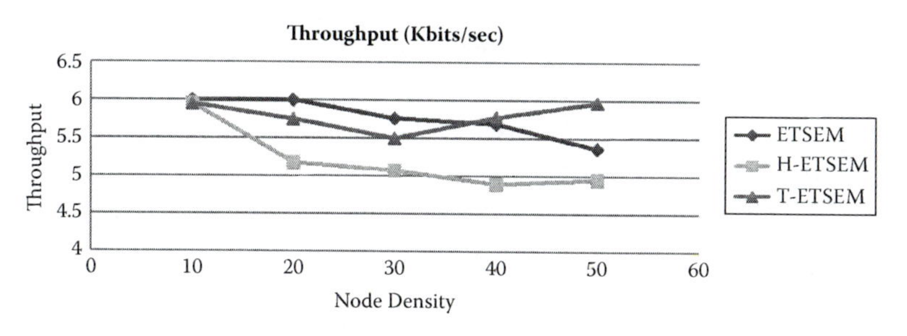

Figure 5.11 Average throughput vs. node density. (From M. Chakraborty and N. Chaki, 2012. *Lecture Notes in Computer Science*. With permission.)

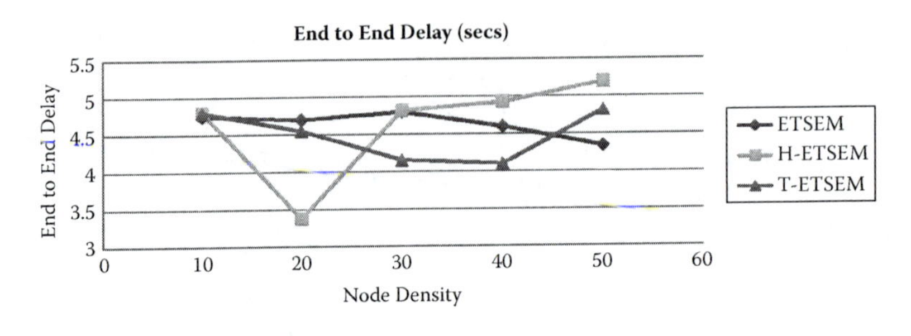

Figure 5.12 Average end-to-end delay vs. node density. (From M. Chakraborty and N. Chaki, 2012. *Lecture Notes in Computer Science*. With permission.)

updating. With TIDS being more efficient than HIDS, T-ETSeM has fewer delay values than H-ETSeM.

5.5.2.4 Jitter In the context of computer networks, the term *jitter* is often used as a measure of the variability over time of the packet latency across a network. A network with constant latency has no variation (or jitter). Packet jitter is expressed as an average of the deviation from the network mean latency. It is an important QoS factor in the assessment of network performance.

Figure 5.13 maps the jitter values for ETSeM, H-ETSeM, and T-ETSeM. Jitter for ETSeM seems to increase with an increase in the number of nodes. However, jitter values for H-ETSeM and T-ETSeM are very similar. Although initially high, these values tend to decrease and remain much lower than that of ETSeM alone.

5.5.2.5 Average Energy Consumption Energy consumption of the nodes is the most critical performance parameter for many energy-constrained networks such as MANETs and sensor networks. The energy values are not obtained from any QualNet stats. Energy for

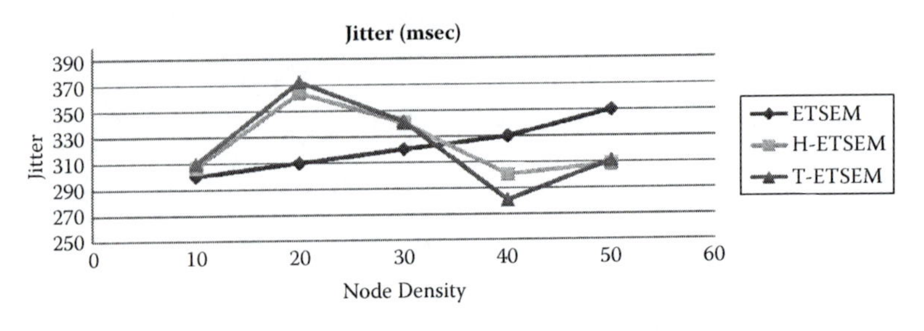

Figure 5.13 Jitter vs. node density. (From M. Chakraborty and N. Chaki, 2012. *Lecture Notes in Computer Science*. With permission.)

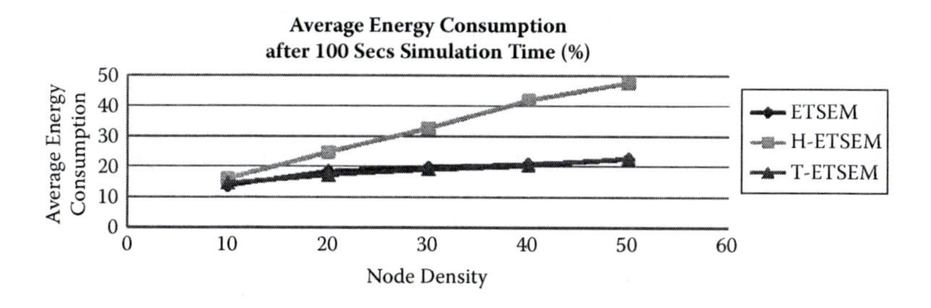

Figure 5.14 Average energy consumption vs. node density. (From M. Chakraborty and N. Chaki, 2012. *Lecture Notes in Computer Science.* With permission.)

nodes, their expense for packet forwarding and packet processing, and their updating has been additionally coded into the ETSeM code. While incorporating HIDS and TIDS, the energy values of nodes have been accordingly updated. Figure 5.14 shows the average energy consumption for all nodes in the network. The results are very interesting and informative. It is expected that ETSeM will have the least energy consumption compared to H-ETSeM and T-ETSeM. This is because H-ETSeM and T-ETSeM do additional packet processing compared to ETSeM for the *h-rate* and trust evaluation processes. This is indeed the case.

However, the average energy consumption of nodes for ETSeM and T-ETSeM is almost same. The lines in the graph are almost identical and overlapping. This implies that TIDS is very lightweight and does not affect the energy consumption of nodes. The energy line for H-ETSEM is much steeper and increases linearly with increase in the number of nodes. This establishes the fact that TIDS is always a better choice than HIDS.

5.6 Conclusion

A survey of the state of the art in multipath routing was done in Section 5.2 of this chapter. The survey reflects that there exist several algorithms that are energy efficient. However, none of these routing schemes uses the trustworthiness of intermediate nodes to decide the degree of the multipath route selection process. This is the novelty of E^{2}TSeM. Not only has security been incorporated as a parameter for route selection process, but two different approaches have been separately implemented and simulated. The simulation results establish that TIDS incorporated with ETSeM provides an efficient modification of ETSeM in considering the trust values of nodes in the route selection process. Hence, ETSeM incorporated with TIDS may be proposed as the new routing protocol, named extended energy-aware, trust-based, selective multipath routing protocol (E^2TSeM), for wireless ad-hoc mobile networks.

References

1. Z. Liang and W. Shi, PET: A Personalized Trust Model with Reputation and Risk Evaluation for P2P Resource Sharing, in *Proceedings of the 38th Annual Hawaii International Conference on System Sciences (HICSS 2005)*, track 7, vol. 7, p. 201b.
2. J.H. Cho, A. Swami, and I.R. Chen, Modeling and Analysis of Trust Management with Trust Chain Optimization in Mobile Ad hoc Networks, in *Proceedings of the 2009 International Conference on Computational Science and Engineering*, vol. 2, pp. 641–650.
3. L. Xiong and L. Liu, Building Trust in Decentralized Peer-to-Peer Electronic Communities, at *Proceedings of 5th International Conference on Electronic Commerce Research, ICECR-5*.
4. W.G. Wang, M. Mokhta, and M. Linda, C-Index: Trust Depth, Trust Breadth, and a Collective Trust Measurement, in *Proceedings of the Hypertext 2008 Workshop on Collaboration and Collective Intelligence*, pp. 13–16.
5. J. Luo, H.X. Liu, and M.Y. Fan, A Trust Model Based on Fuzzy Recommendation for Mobile Ad-Hoc Networks, *Computer Networks*, 53(14), 2396–2407, 2009.
6. S. De, C. Qiao, and H. Wu, Meshed Multipath Routing with Selective Forwarding: An Efficient Strategy in Wireless Sensor Networks, *Elsevier Computer Networks: Sepcial Issue on Wireless Sensor Networks*, 43(4), 481–497, 2003.
7. Jie Zhang, Choong Kyo Jeong, Goo Yeon Lee, and Hwa Jong Kim, Cluster-Based Multi-Path Routing Algorithm for Multi-Hop Wireless Network, *International Journal of Future Generation Communication and Networking*, 1(1), 67–74, 2009.
8. S. Mueller, R.P. Tsang, and D. Ghosal, Multipath Routing in Mobile Ad hoc Networks: Issues and Challenges, in *MASCOTS 2003*, ed. M.C. Calzarossa and E. Gelenbe, LNCS, vol. 2965, Springer, Berlin, 2004, pp. 209–234.
9. M. Radi, B. Dezfouli, S.A. Razak, and K.A. Bakar, LIEMRO: A Low-Interference Energy-Efficient Multipath Routing Protocol for Improving QoS in Event-Based Wireless Sensor Networks, in *Fourth International Conference on Sensor Technologies and Applications, SENSORCOMM*, 2010, pp. 551–557.
10. Siddiqui Muhammad Shoaib, Amin Syed Obaid, Kim Jin Ho, and Hong Choong Seon, MHRP: A Secure Multi-Path Hybrid Routing Protocol for Wireless Mesh Network, in *Military Communications Conference, MILCOM 2007*, IEEE, pp. 1–7.

11. Zhiyuan Li and Ruchuan Wang, A Multipath Routing Algorithm Based on Traffic Prediction in Wireless Mesh Networks, *Communications and Network*, 1(2), 82–90, 2009.
12. Ye Ming Lu and Vincent W.S. Wong, An Energy-Efficient Multipath Routing Protocol for Wireless Sensor Networks, *International Journal of Communication Systems*, 20(7), 747–766, 2007.
13. J. Agrakhed, G.S. Biradar, and V.D. Mytri, Energy Efficient Interference Aware Multipath Routing Protocol in WMSN, in *Annual IEEE India Conference (INDICON)*, 2011, pp. 1–4.
14. Moufida Maimour, Maximally Radio-Disjoint Multipath Routing for Wireless Multimedia Sensor Networks, in *Proceedings of the 4th ACM Workshop on Wireless Multimedia Networking and Performance Modeling*, 2008, pp. 26–31.
15. R. Chaki and N. Chaki, IDSX: A Cluster Based Collaborative Intrusion Detection Algorithm for Mobile Ad-Hoc Network, in *International Conference on Computer Information System and Industrial Management Applications (CISIM'07)*, Minneapolis, MN, June 28–30, 2007, pp. 179–184.
16. Aikaterini Mitrokotsa, Nikos Komninos, and Christos Douligeris, Intrusion Detection with Neural Networks and Watermarking Techniques for MANET, in *Proceedings of the IEEE International Conference on Pervasive Services*, 2007, pp. 118–127.
17. Noman Mohammed, Hadi Otrok, Lingyu Wang, Mourad Debbabi, and Prabir Bhattacharya, Mechanism Design-Based Secure Leader Election Model for Intrusion Detection in MANET, *IEEE Transactions on Dependable and Secure Computing*, 8(1):89–103, 2011.
18. H. Yang, J. Shu, X. Meng, and S. Lu, SCAN: Self-Organized Network-Layer Security in Mobile Ad hoc Networks, *IEEE Journal on Selected Areas in Communications*, 24, 261–273, 2006.
19. Debdutta Barman Roy, Rituparna Chaki, and Nabendu Chaki, BHIDS: A New Cluster Based Algorithm for Black Hole IDS, *Journal on Security and Communication Networks*, 3(2–3), 278–288, 2010.
20. Manali Chakraborty and Nabendu Chaki, ETSeM: An Energy-Aware, Trust-Based, Selective Multi-Path Routing Protocol. *Computer Information Systems and Industrial Management: Lecture Notes in Computer Science*, (7564):351–360, 2012.
21. Poly Sen, Nabendu Chaki, and Rituparna Chaki, HIDS: Honesty-Rate Based Collaborative Intrusion Detection System for Mobile Ad-Hoc Networks, *Computer Information Systems and Industrial Management*, 2008, pp. 121–126.
22. Novarun Deb and Nabendu Chaki, TIDS: Trust-Based Intrusion Detection System for Wireless Ad-hoc Networks. *Computer Information Systems and Industrial Management: Lecture Notes in Computer Science*, (7564):80–91, 2012.

6

BLACKHOLE ATTACK DETECTION TECHNIQUE

DEBDUTTA BARMAN ROY
AND RITUPARNA CHAKI

Contents

6.1 Introduction

A mobile ad-hoc network (MANET) is formed by a group of mobile wireless nodes without the assistance of a fixed network infrastructure [17]. The nodes in a MANET cooperatively forward packets so that the nodes beyond the radio ranges can communicate with each other. These properties make MANETs immensely important as flexible network platforms. With so much flexibility comes the problem of security. The open medium, dynamically changing network topology, cooperative algorithm, lack of centralized monitoring [3, 9, 13], etc., make MANETs highly vulnerable to different types of attacks.

There may be active attacks or passive attacks in a MANET [24]. Active attacks can be of many types, such as blackhole, routing loops, network partitioning, cache poisoning, selfishness, etc. Each attack type requires specialized attention, extensive evidence gathering, and comprehensive analysis, as there is often little difference in intrusion and legitimate operation.

In a blackhole attack, a malicious node impersonates a destination node by sending a spoofed route reply packet to a source node that initiates a route discovery. By doing this, the malicious node can deprive the traffic from the source node. A blackhole has two properties. First, the node exploits the ad-hoc routing protocol, such as the ad-hoc on-demand distance vector (AODV), to advertise itself as having a valid route to a destination node, even though the route is spurious, with the intention of intercepting packets. Second, the node consumes the intercepted packets. A malicious node always responds positively with a RREP message to every RREQ, even though it does not really have a valid route to the destination node. Since a blackhole does not have to check its routing table, it is the first to respond to the RREQ in most cases. When the data packets routed by the source node reach the blackhole node, it drops the packets rather than forwarding them to the destination.

The goal of intrusion detection is seemingly simple: to detect attacks. However, the task is difficult, and in fact, intrusion detection systems do not detect intrusions at all—they only identify evidence of intrusions, either while they are in progress or after the intrusion [14–16]. For accurate intrusion detection, we must have reliable and complete data about the target system's activities. Reliable data collection is a complex issue in itself [18].

AODV [28] is a reactive routing protocol where the network generates routes at the start of communication. Each node has its own sequence number, and this number increases when connections change. Each node selects most of the recent channel information, depending on the largest sequence number.

AODV routing methodology is quite susceptible for blackhole [29] attacks. A malicious node advertises itself as having a valid route to a destination node, even though the route is spurious, with the intention

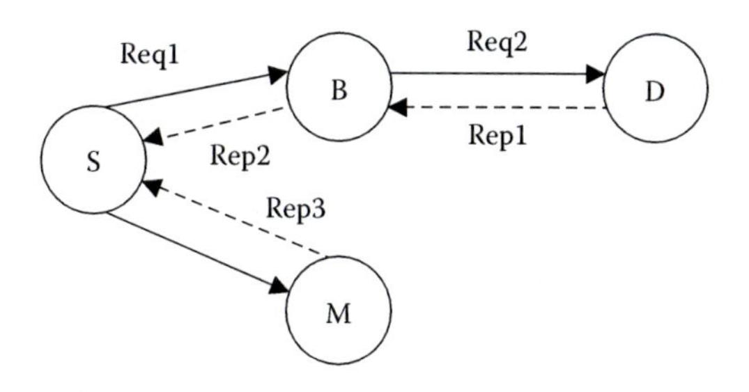

Figure 6.1 Blackhole attack.

Table 6.1 Values of RREQ and RREP

	RREQ		RREP		
	REQ1	REQ2	REP1	REP2	REP3
Intermediate node	S	B	D	B	M
Destination node	D		D		D
Source node	S				
Dst_Seq	11		12		20

of intercepting packets. In AODV, a destination sequence (Dst_Seq) number is used to determine the freshness of routing information contained in the message from the originating node. When generating a RREP message, a destination node compares its current sequence number and Dst_Seq in the RREQ packet plus one, and then selects the larger one as RREP's Dst_Seq. Upon receiving a number of RREPs, a source node selects the one with greatest Dst_Seq in order to construct a route. To succeed in the blackhole attack, the attacker must generate its RREP with Dst_Seq greater than the Dst_Seq of the destination node. It is possible for the attacker to find out the Dst_Seq of the destination node from the RREQ packet. In general, the attacker can set the value of its RREP's Dst_Seq based on the received RREQ's Dst_Seq. However, this Dst_Seq may not be present in the current Dst_Seq of the destination node. Figure 6.1 shows an example of the blackhole attack. The values of RREQ and RREP used in the attack are shown in Table 6.1.

In Table 6.1, the intermediate node indicates a node that generates or forwards a RREQ or RREP packet. The source node indicates the nodes that generate and send packets meant for destination nodes. Here, it is assumed that no other node has any routing information to the destination node D at this point in time. The source node S constructs a route in order to communicate with destination node D. Let's assume that source node S has a value of 11 as Dst-Seq for the destination node D. Hence, source node S sets its RREQ_(Req1) with this Dst_Seq value and broadcasts the same. Upon receiving RREQ_(Req1), the intermediate node B forwards RREQ_(Req2), since it is not the destination node. Let's assume that to impersonate the destination node, the attacker M sends spoofed RREP(Rep3), shown in Table 6.1, with the intermediate source node, the destination node the same with D, and increased Dst_Seq (in this case by 20) to the source node S. At the same time, the destination node D, which received RREQ_(Req2), sends RREP (Rep1) with Dst_Seq incremented by one to node S. Although the source node S receives two RREP, based on Dst_Seq the RREP(Rep3) from the attacker M is judged to be the most recent routing information and the route to node M is established. M, an intruder node, receives packets from the source and destroys them instead of forwarding them to destination node D.

6.2 State of the Art in Wireless Ad-Hoc Networks

A brief study on intrusion detection systems (IDSs) is presented in this chapter. Intrusion prevention measures, such as authentication and encryption [16, 20, 26], can be used as the first line of defense against attacks in MANET. Secure Pebblenets, proposed in [21], provide a distributed key management system based on symmetric encryption. However, even if these prevention schemes can be implemented perfectly, they still cannot eliminate all attacks, especially the internal or insider attacks. For example, mobile nodes (and their users) can be captured and compromised. The attacker can then obtain the cryptographic keys. There are many other internal attack methods, including using worms and viruses, that propagate within MANETs.

Structural and behavioral differences between wired and wireless mobile networks make existing IDS designs inapplicable to the wireless networks. As discussed above, wireless network communications are conducted in open-air environments. Thus, network monitoring in wireless ad-hoc networks is performed at every network node [7, 8, 27]. This approach is inefficient due to network bandwidth consumption and increased computations resources that are highly limited in a wireless network. Host-based monitoring also contributes to a high amount of processing on each host, shortening battery life and slowing down the host. Physical mobile host security is an issue, as each host contains keys used to encrypt information over the network, and if captured, the network is subject to eavesdropping.

IDSX [4] is another collaborative algorithm that offers an extended architecture and is compatible with heterogeneous IDS already deployed in the participating nodes. In the high level of the architecture of the IDSX mechanism, the cluster heads act as the links across different clusters. The cluster heads are IDSX enabled, and hence can utilize alerts to generate the alarms. Alerts represent the potential security breaches as identified by local IDS active nodes. The IDSX nodes are authorized to make the final decision of discarding a node forever after aggregating and correlating the alerts that have been generated over a predefined period of time.

Applying functionality-based network IDS models also has limitations. An anomaly detection model is built on long-term monitoring and classifying of what is normal system behavior. Ad-hoc wireless networks are very dynamic in structure, giving rise to apparently random communication patterns, thus making it challenging to build a reliable behavioral model. Misuse detection requires maintenance of an extensive database of attack signatures, which in the case of ad-hoc networks would have to be replicated among all the hosts. A few papers have suggested IDSs targeted at wireless networks.

In order to overcome the drawbacks of central analysis techniques, distributed analysis started to evolve. Several agent-based distributed IDSs have been developed in recent years, and all adopted distributed analysis techniques [6, 19, 23, 25, 36]. Autonomous agents for intrusion detection (AAFID) [5] is a distributed anomaly detection system that employs autonomous agents at the lowest level for data collection and analysis. At the higher levels of the hierarchy, transceivers and monitors are used to obtain a global view of activities. The mobile agent intrusion detection system (MAIDS) [22] developed the agent that transfers different types of data to another agent, which has the capability to analyze certain data types.

In [27], a distributed IDS with a cooperative decision algorithm is presented. Reference [30] addresses one aspect of the problem of defending MANET against computer attacks. This approach is based on distributing an anomaly-based intrusion detection system. This uses a three-level hierarchical system. In layer 1 local IDSs are attached to every node in the MANET, collecting raw data of network operation and computing a local anomaly index measuring the difference between current node operation and the baseline of normal operation. In layer 2 the cluster head with the lowest node ID fuses node indexes, producing a cluster level anomaly index, and periodically broadcasts to the manager. It also broadcasts a message requesting its neighboring nodes to join the cluster. Since this architecture relies on the use of mobile agents, it increases computational complexity in creating and managing all agents.

In another significant work, a solution has been offered that combines the advantage of agent-based distributed analysis and a clustering-based intrusion detection technique with balanced data. The advantage of MAID is that it has the capability to analyze certain data types and collect and transfer different types of data. One or more mobile nodes keep watch on the activities of all nodes and report intrusions on their own or make decisions collaboratively [4]. The agent collects a huge amount of raw network packets from different sources while the process is running. Then data instances are normalized to standard form for solving the problem that different features are on different scales. While the process is running, if the central IDS loses communication with the agent ID, other central IDSs will take responsibility, to avoid the problem of single-point failure.

One of the possible attacks in a MANET is a blackhole attack, where a malicious node includes itself in routes, and then simply drops packets in spite of forwarding. Another motivation for dropping packets in self-organized MANETs is resource preservation. Some solutions for detecting and isolating packet droppers have been recently proposed [5]. However, almost all of them employ the promiscuous mode monitoring approach (watchdog). The promiscuous mode assumes an absolutely trusted environment, which often is not the reality. Besides, it has an adverse impact on the energy efficiency of nodes. In [5], a monitoring algorithm has been proposed for overcoming some of the watchdog's shortcomings, and improves the efficiency in detection. To overcome false detections due to nodes' mobility and channel conditions, the Bayesian technique is proposed for the judgment, allowing node redemption before judgment. Finally, a social based approach for the detection approval and isolation of malicious nodes has been suggested.

Work in [11] suggests a method of observing the packet flow at each node. A total of 141 traffic-related and topology-related features have been defined. In [12], an extended finite state automaton has been defined according to the specification of AODV. Both of these approaches use static training data to define the normal state. As the topology of MANETs changes at very short intervals, the static training data are insufficient to capture the states of the network. An algorithm based on the dynamic training method is proposed in [3], where training data are updated at regular intervals.

In [29], the effect of a blackhole attack is studied with variable numbers of connections to and from the destination nodes. They consider that all nodes behave normally and are closely placed. The node that behaves abnormally is placed in a scattered way. As this is a mobile ad-hoc network, this phenomenon may not occur frequently. The nodes are dynamically trained by data computed using an equation. In each time interval ΔT, the nodes are trained with new data. This forms an overhead for each node.

In [31], an approach called PC has been proposed for preventing blackhole attacks when more than one node behaves maliciously. The authors used a "fidelity table," wherein every participating node is assigned a fidelity level that acts as a measure of reliability of that node. In case the level of any node drops to 0, it is considered to be a malicious node, termed a blackhole, and is eliminated. The source node transmits the RREQ to all its neighbors. Then the source waits for TIMER seconds to collect the replies, RREP. A reply is chosen based on the following criteria: In each of the received RREP, the fidelity level of the responding node and each of its next-hop levels are checked. If two or more routes seem to have the same fidelity level, they select the one with the least hop count; otherwise, select the one with the highest level. The fidelity levels of the participating nodes are updated based on their faithful participation in the network. On receiving the data packets, the destination node will send an acknowledgment to the source, whereby the intermediate node's level will be incremented. If no acknowledgment is received, the intermediate node's level will be decremented. Here, for each node being maintained, the fidelity table is an overhead that may reduce the performance level of the network.

A hierarchical secure routing protocol for detecting and defending blackhole attacks (HSRBH) is proposed in [32]. The proposed HSRBH is an on-demand routing protocol, and it uses only symmetric key cryptography to discover a safe route against blackhole attacks. The source node initiates RREQ and sends it to the sink node. When the sink node and other intermediate group leaders having a fresh enough route receive RREQ, they generate RREP, which contains a message authentication code (MAC) that is calculated using an intergroup shared key. For each RREP two-step verification is done. This approach is secure enough, but it increases the processing overhead by two steps for verification.

In paper [33], a new methodology called SRSN has been proposed. SRSN is based on the strict increment of sequence number of RREQ packet combined with reliable end-to-end acknowledgment to detect false route information. SRSN makes a little edition of the Dynamic Source Routing (DSR) protocol. It can defend against blackhole attacks without increasing the consumption of the resource and payload too much. The main idea of this algorithm is that nodes in an ad-hoc network receive RREQ packets from other nodes. Then it checks the sequence number to decide whether the route record can be trusted. If the sequence number is continuous, it means that the route record is true. Thus, it can be added to the trusted routing list and used in data delivery. Otherwise, the route record is in a suspicious status and added to the suspicious routing list. This approach takes more time during route generation, which degrades the overall performance of the network.

SAODV, a secure routing protocol based on AODV, has been proposed in [34]. The SAODV algorithm claims to be able to avoid blackhole attacks. To reduce the chances of attack, this algorithm proposes to wait and check the replies from the entire neighboring node. The source node will store the sequence number and the time at which the packet arrives in a collect route reply table (CRRT). If more than one path exists in CRRT, then it randomly chooses a path from CRRT. This reduces the chances of blackhole attacks. Here, for each node being maintained, a CRRT is an overhead that may reduce the performance level of the network.

Work in [35] suggests implementing a new protocol called IDSAODV. In this approach, the first RREP message is used to initiate DATA transfer. Whenever the node gets a second RREP message for the same destination, a new route is used. The overhead of making a trust table and time delay is overcome here. This approach is not suitable if the malicious node is smart enough. That means if the malicious node sends RREP after a period of time, then the source does not understand which route is to be followed.

In [36] the model presents a secure communication between the mobile nodes. A scenario of data transmission between the two mobile nodes has been considered. Whenever a source wants to transmit the data packets to the destination, it ensures that the source is communicating with a real node via the cluster head. The authentication service uses key management to retrieve the public key, which is trusted by the third party for identification of the destination. The destination also uses a similar method to authenticate the source. After execution of the key management module, a session key is invoked; this is used by both the source and the destination for further communication confidentially. In this way, all the important messages are transmitted to the destination.

The study of the state of the art shows that MANETs are highly susceptible to blackhole attacks [29]. Detection methodologies are proposed based on the fidelity table [31], dynamic learning [29], Bayesian techniques [5], etc. Frequent updating of the fidelity table leads to serious overhead. Dynamic learning and the Bayesian approach need regular updating of training data. It is observed that the vast majority of the existing works for handling blackhole attacks suffer from multiple limitations. These include overhead due to communication [31], overhead due to data structure and table maintenance [31, 29], performance and operational bottlenecks on the cluster head [32], poor node mobility [31], and correctness [35, 36] of the algorithm in terms of both false intrusion detection and failure to detect an attack.

6.3 Description of the Processes

This chapter discusses a new cluster-based blackhole detection approach. As in any cluster-based approach, the cluster formation and cluster head selection are the basic processes of this method. In this section, details of the cluster formation, updating, and cluster head selection are described for the blackhole detection technique.

6.3.1 *Assumptions*

The following are the assumptions that we have made in order to design the proposed algorithm:

1. Every node has a unique ID in the network, which is assigned to a new node collaboratively by existing nodes.
2. The nodes in the network form a two-layered cluster. A cluster head at the inner layer is represented as $CH_{(1,i)}$, where 1 signifies inner layer and i stands for the cluster number.
3. Each cluster is monitored by only one cluster head.
4. The cluster membership is restricted up to one hop.
5. The cluster head should not be the malicious node.
6. The node with minimum node ID in a cluster becomes the cluster head for that cluster.
7. Each cluster has its own cluster ID that is given by the cluster head.
8. The cluster head at layer 2 only communicates with the cluster head at layer 1.
9. The cluster head at layer 1 only monitors the cluster member; it does not act as an intermediate node.

6.3.2 *Layered Architecture*

In this section the layered approach is introduced to reduce the load of processing on each cluster head. From a security point of view, this will also reduce the risk of a cluster head being compromised.

The entire network is divided into clusters as shown in Figure 6.2. The clusters may be overlapped or disjoint. Each cluster has its own cluster head and a number of nodes designated as member nodes. Member nodes pass on the information only to the cluster head. The cluster head is responsible for passing on the aggregate information to all its members. The cluster head is elected dynamically and maintains the routing information.

GN is the guard node, used for monitoring the malicious activity. The main purpose of the guard node is to guard the cluster from possible attacks. The guard node has the power to monitor the activity of any node within the cluster. The guard node reports to the cluster head of the respective layer in case a malicious activity is detected. A

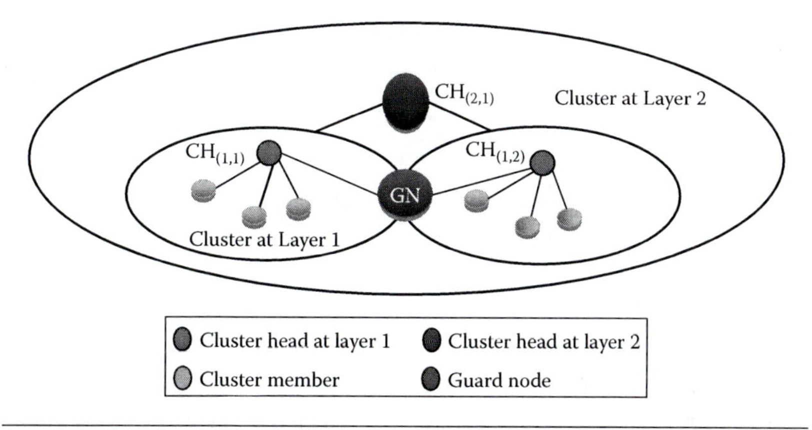

Figure 6.2 The layered structure. (From Barman, Roy D. et al. 2009. *Security and Communication Networks*, John Wiley & Sons, New York. With permission.)

cluster head in the inner layer ($CH_{1,i}$) detects a malicious activity and informs the cluster head CH_2 of the outer layer to take appropriate action. It's the duty of ($CH_{1,i}$) to check the number of false routes generated by any node. The cluster head CH_2 of the outer layer takes upon itself the responsibility of informing all nodes of the inner layer about the malicious node. Often, a cluster is defined with the cluster head and the nodes within one hop distance from the cluster [4].

Cluster head: A cluster head as defined in the literature, serves as a local coordinator for its cluster. The cluster head has information about each member of the cluster. It provides a unique node ID for each cluster member. The cluster head does not work as an intermediate node. It has only the responsibility to maintain the cluster and inform the cluster head in layer 2 of any malicious attack in the cluster.

Cluster member: A cluster member is a node that is not the cluster head. It has the neighbor list of only its one-hop neighbor, including the cluster head.

The responsibility of intrusion detection is shared among nodes in the cluster. Individual nodes within the inner cluster collect the raw data from the network to detect any intrusion, and then send the information to the cluster head at layer 1 [10]. The duty of a cluster head in the inner layer (CH_1) is to detect a malicious activity and inform the cluster head CH_2 of the outer layer to take appropriate action. The cluster head CH_2 of the outer layer takes upon itself the responsibility of informing all nodes of the inner layer about the malicious node. The cluster head at layer 1 consists of detailed routing information of each node in the cluster. It's the duty of CH_1 to check the number of false routes generated by any node.

6.3.3 Cluster Head Selection

Cluster head selection is one of the essential aspects in a clustering algorithm. TACA is a distributed algorithm that takes into account the mobility of a node and its available battery power as the parameters to decide its suitability as a cluster head. A large mobility factor indicates a slower node, and a small mobility factor indicates a faster node. Available battery power is the energy contained in the node at the instant of weight calculation. These two parameters are added with different weight factors to find the weights of the individual nodes [1].

The weighted clustering algorithm (WCA) obtains one-hop clusters with one cluster head. The election of the cluster head is based on the weight of each node. It takes four factors into consideration and makes selection of the cluster head and maintenance of the cluster more reasonable. The four factors are node degree (number of neighbors), distance summation to all its neighboring nodes, mobility, and remaining battery power [3]. Although WCA has shown better performance than all the previous algorithms, it has a drawback in knowing the weights of all the nodes before starting the clustering process, and in draining the CHs rapidly. As a result, the overhead induced by WCA is very high.

In a modified weighted clustering algorithm [2], the cluster head selection is based on four parameters: cluster size, distance of the cluster from its neighbors, mobility of the node, and battery power of the node. The node with minimum weight is chosen as the cluster head. Whenever a new node arrives in the cluster, the node first computes its weight and then becomes a member of the cluster. If there is no cluster, then the new node forms a cluster and becomes the cluster head of that cluster [1].

In the existing literature, the trust values of nodes are not often taken as key factors for cluster head selection. However, in the proposed method described in this section, the trust value of a node is considered one of the key parameters in selecting the cluster head.

In opinion-based trust evaluation models [37], each node is responsible for evaluating a node's behavior and categorizing it as a well-behaved, misbehaving, or suspect node. Every node is responsible for monitoring the behavior of its neighbor and then discriminating misbehaving nodes from well-behaving nodes. Each node passively receives a lot of information about the network. This information is used to build trust levels for different nodes.

In the proposed key management scheme, whenever a node is needed to initiate route discovery, it constructs RREQ packets and generates SMSG (start message). SMSG consists of the secret key that has to be shared between the source and the destination, and the digital signature of the same. The source node now forwards the RREQ along with the SMSG. Once the destination receives the RREQ along with SMSGs, it verifies the digital signature via polling. It chooses the shared secret key that has been proved to be valid and detects the misbehavior if the digital signature sent via a path seems to be invalid. The destination reports the source regarding the misbehavior, and hence every intermediate node records it for the future trust factor calculation. Once the source receives the RREP, it starts transmitting the data encrypted via a keyed HMAC algorithm using the secret key shared between the source and destination.

This layered architecture helps to reduce the processing overhead of the cluster head at layer 1 at the cost of a slight increase in the communications overhead.

6.3.3.1 Cluster Head Selection Algorithm The mobile ad-hoc network can be modeled as a unidirectional graph $G = (V, L)$, where V is the set of mobile nodes and L is the set of links that exist between the nodes. We assume that there exists a bidirectional link L between the nodes, and when the distance between the nodes $d_{i,j} < t_{range}$ (transmission range), then $L_{i,j}$ can be a link between nodes i and j.

Mobility factor: Let δ be the maximum permissible speed of any network. The average of last n displacements gives the average speed of any node. Thus, the difference of average speed (S_{av}) finds the mobility factor (ΔM) of a node:

$$\text{Mobility factor } (\Delta \text{M}) = \delta - S_{av} \tag{6.1}$$

A large mobility factor indicates a slower node, and a small mobility factor indicates a faster node.

Available battery power (P_{av}): Let P_{DT}, P_{DR}, and P_{DR} denote the battery power consumed by a node during data transmission, data received, and overhearing of data in the network. Thus, the available battery power (P_{av}) of a node can be the difference between the total battery power of a node and the battery power consumed ($P_{consume}$) by the node.

$$P_{consume} = P_{DT} + P_{DR} + P_{DR} \tag{6.2}$$

$$P_{av} = P_{total} - P_{consume}$$

Average trust rate ($TR_{(i,CH)}$): Let $T_{NN}{}^{CH}$ be the trust value given by a cluster head to the neighbor node and O_i^{NN} be the opinion given by the neighbor node (NN) to the node i. Thus, the average trust rate of each node is calculated to know the reliability of the node in the network. The average trust rate of each node is computed as

$$TR_{(i,CH)} = \Sigma T_{NN}{}^{CH} * O_i^{NN} / \Sigma NN \tag{6.3}$$

6.3.4 Description of the Proposed IDS

In this section we present the outline for the proposed solution [4]. Each route request packet here contains a destination sequence number. The corresponding RREP quotes this number. A source node judges the packet with the highest destination sequence number as the one containing the most recent routing information when it receives multiple RREP messages. When a node in the inner layer identifies a blackhole attack within the cluster, it informs $CH_{(1,i)}$. $CH_{(1,i)}$ informs CH_2 about the malicious node. The cluster head at layer 2 broadcasts this information to all cluster heads at layer 1. The layer 1 cluster heads inform the respective cluster members. In Figure 6.3, node A broadcasts RREQ (Route REQUEST packet) to all the members for destination node G.

The intermediate node B sends another RREQ2 to node D. If the attacker node M gets the RREQ, then it sends the RREP to the

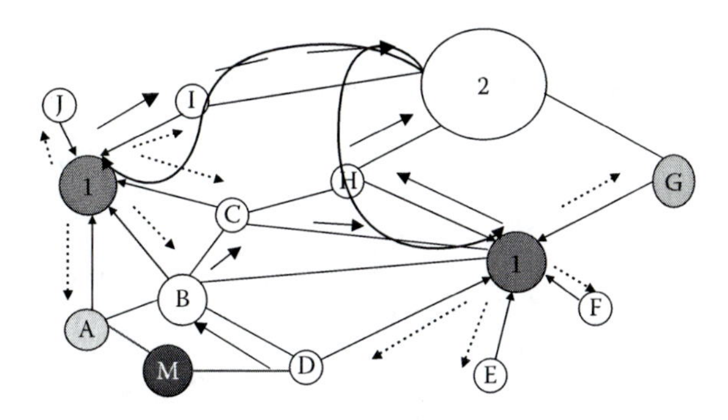

Figure 6.3 Blackhole detection. (From Barman, Roy D. et al. 2009. *Security and Communication Networks*, John Wiley & Sons, New York. With permission.)

source node A with a larger destination sequence number. The node A modifies its route as given by M. However, M has no route to G. Thus, M sends the packets meant for G to one of its neighbors, say D. Now, according to BHIDS, node D identifies this problem, alerts the corresponding cluster head of layer 1 about the problem. The cluster head at layer 1 informs the cluster head at layer 2 and accordingly takes action.

In this algorithm the false packet rate (FPR) is computed with respect to some threshold value:

$$FPR = \frac{R_i - F_i}{R_i} \times 100\% \tag{6.4}$$

Here, $R_i \geq$ number of packets received by intermediate node N_i from a particular source, and $F_i \geq$ number of packets forwarded by node N_i to the destination.

The process of blackhole detection has been described in a few simple steps below.

6.3.5 Algorithm for Cluster Head Selection

6.3.5.1 Steps in Calculating Weight

Step 1: Begin.

Step 2: For every node $v \varepsilon V$, compute the average speed S_{av} – D_t/n, where D_t implies total distance covered in nsec.

Step 3: Compute mobility $\Delta M = \delta - S_{av}$.
Step 4: Compute available battery power:

$$P_{av} = P_{total} - P_{consume}$$

where P_{av} implies available battery power.
Step 5: Compute average trust rate ($TR(i,CH)$) using Equation (6.3).
Step 6: Compute weight of node v, $W_v = W_1\Delta M + W_2 P_{av} + W_3 TR_{av}$.
Step 7: End.

Once the weight is calculated, the cluster head selection procedure is started.

6.3.5.2 *Algorithm for Cluster Head Selection*

Step 1: Begin.
Step 2: For every $v \varepsilon V$, if $W_v > W_i$, where i implies the neighbor node of v, the

$$\text{HEAD} = v$$

Step 3: For every $x \varepsilon V_{uncover}$, if $dist_{(head,x)} < head_{range}$, then set

$$head_x = head$$

Step 4: End.

6.3.5.3 *Cluster Head Updating*

Step 1: Begin.
Step 2: Verify the threshold on the cluster head battery power. If ($P_{av} < P_{th}$), the cluster head sends a Life_Down message to all its neighbors.
Step 3: Verify the threshold on the cluster head trust value. If ($TR_{av} < TR_{th}$), the cluster head sends a Trust_Loss message to all its neighbors.
Step 4: All the nodes in the cluster then participate in the reelection of cluster head.
Step 5: The node with maximum weight is selected as the cluster head.
Step 6. End.

6.3.6 Algorithm for Blackhole Detection

This section describes the algorithm of blackhole detection using a cluster-based approach.

Step 1: Begin.
Step 2: The source node broadcasts a route request message RREQ along with the destination node ID.
Step 3: The destination node replies with route reply message RREP, which contains the source ID and a Dst_seq number.
Step 4: On receipt of the route reply, the source node checks the message with the highest Dst_seq of RREP.
Step 5: The source node updates the routing table as given by the destination node D.
Step 6: The source node S chooses the next-hop node N_i from the route table sent by the RREP message by destination node D and sends a packet for the destination node to that node.
Step 7: Node N_i captures the packet.
Step 8: It observes the destination address D and destination sequence number S_N from the packet.
Step 9: If $D = N_i$, then:

```
        Receive packet and increment Ri by 1
        Terminate the detection process
Else go to step 4
Endif
```

Step 10: If a route to D is found in the routing table, then:

```
        Send the packet to the next hop node
        specified in route table
  Increment Fi by 1 and Ri by 1
Endif
```

Step 11: The node calculates FPR using formula (6.1).
Step 12: If FPR is greater than the threshold value, then:

```
        Inform cluster head for Ni at layer 1
  Else go to step 1
Endif
```

Step 13: End.

6.4 Performance Analysis

The metrics used to evaluate the performance are given below:

Throughput of packet forwarding (P_f): The ratio between the number of packets originated by the application layer sources (P_s) and the number of packets received by the sink at the final destination (P_d).

$$P_f = P_d / P_s \tag{6.5}$$

Node mobility (N_m): It signifies how fast the node is changing its position in the network.

$$\text{Pr} \ \alpha \ 1/N_m \tag{6.6}$$

$$\text{Pr} = \text{constant}/N_m$$

Cumulative sum of receiving packet: This is defined as the sequence of the partial sums of all packets received by the destination node.

$$\text{CU}_{\text{Sum}} = \sum_{i=1}^{n} N^{i}_{pkt} \tag{6.7}$$

where n is the total number of packets sent at the ith instance to the destination node.

6.4.1 *Performance Evaluation*

Both NS2 version 2.29.2 with Cygwin-1.5.21 and MATLAB®6 have been used to simulate the proposed BHIDS. The simulation parameters for NS2 simulation are set as shown in Table 6.2.

Due to the collision in the network, the mobility of the node changes frequently. So, we have computed the throughput of packet forwarding with respect to the mobility change and plotted the graph in Figure 6.4. To evaluate the throughput of packet forwarding, simulation is done with 11 nodes with the source node transmitting 100 packets to the destination node. When the network is not under attack, the performance of the network is better. If the network is under a blackhole attack, then the throughput is minimal. This implies

Table 6.2 Simulation Parameters

SIMULATOR	NS2
Number of mobile nodes	10–15
Number of malicious nodes	1
Routing protocol	AODV
Maximum bandwidth	2 Mbps
Traffic	Constant bit rate
Maximum connection	10–50
Maximum speed	10–100 mps
Pause time	5 s

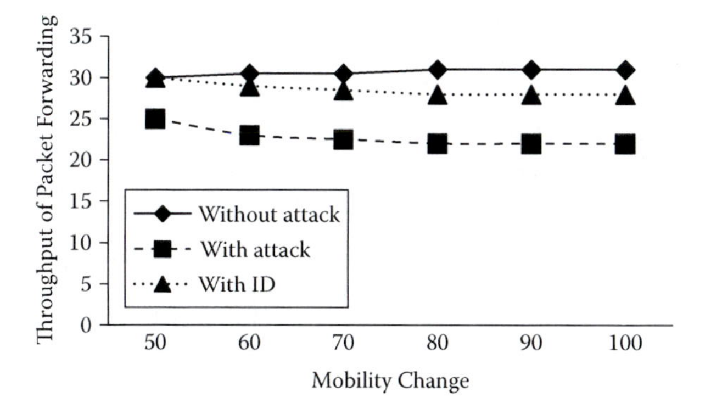

Figure 6.4 The rate of packet drops. (From Barman, Roy D. et al. 2009. *Security and Communication Networks*, John Wiley & Sons, New York. With permission.)

that the destination node does not receive the packets forwarded by the source node. When the network is under attack in the presence of an IDS, then the performance is good, showing that the IDS is able to identify the intruder and the performance of the network becomes better. This series shows that the performance is better initially, then degrades, indicating the presence of an intruder in the network.

In Figure 6.5, it is observed that the performance in the presence of IDS is better than that in the presence of an attacker. This implies that when the network is under a blackhole attack, the intruder drops the packets sent by the source node for the destination node. When the IDS plays its role, it readily identifies the intruder, and the source can securely send all packets to the destination node. Here, as the mobility increases, the performance of the network also increases, showing that at higher mobility, the chances of source and destination nodes coming closer to each other rise. When the source and

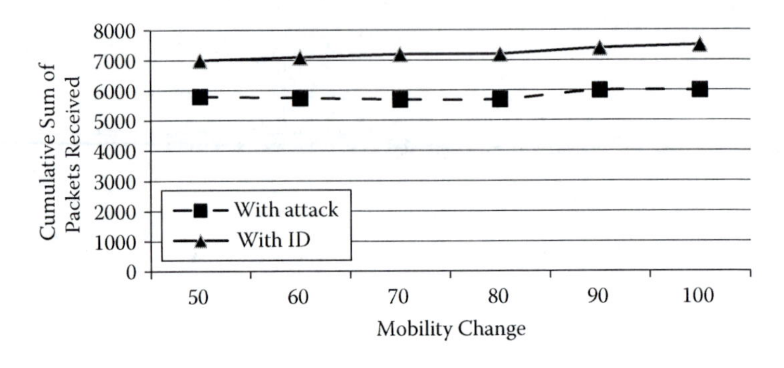

Figure 6.5 Packet delivery ratios. (From Barman, Roy D. et al. 2009. *Security and Communication Networks*, John Wiley & Sons, New York. With permission.)

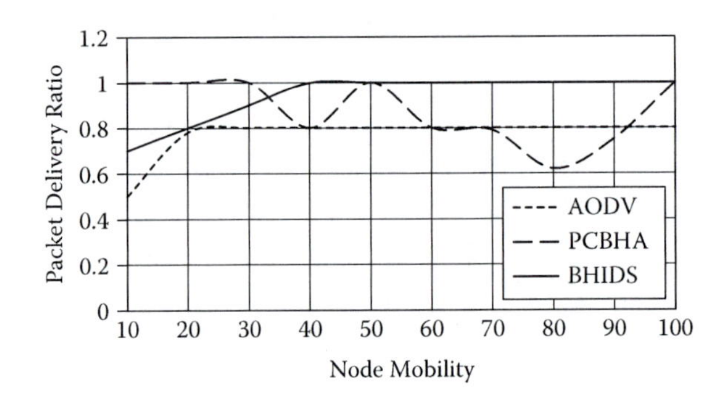

Figure 6.6 Packet delivery ratios. (From Barman, Roy D. et al. 2009. *Security and Communication Networks*, John Wiley & Sons, New York. With permission.)

destination nodes are in the same cluster, the number of packets received is more than when they are in different clusters.

The packet delivery ratio with varying mobility has been presented here. The simulation is done with 11 nodes, with the source node transmitting 100 packets to the destination node. As can be seen from Figure 6.6, with BHIDS the packet delivery ratio is better than AODV and PCBHA. Initially, the packet delivery ratio is lower than with PCBHA, but as the mobility increases, the source node comes closer to the destination node, and the communication overhead among the node decreases, increasing the packet delivery ratio.

6.5 Conclusion

In the existing literature, it is found that communication overhead is one of the major disadvantages for ad-hoc networks. The proposed algorithm BHIDS aims to reduce the communication overhead by making a one-hop cluster. The cluster head of a cluster has direct communication with all its cluster members. The cluster head does not act as an intermediate node while sending data from the source to the destination. The cluster head only monitors the cluster member and communicates with the cluster head at layer 2.

In the proposed BHIDS algorithm, only the neighbor table and route reply table are used. Both tables are updated at the time of route generation. No other tables are required. This reduces the table updating overhead.

As the cluster head only has to monitor the cluster member and communicate with the cluster head at layer 2, the workload of the cluster head is not enough to create a bottleneck. In the proposed BHIDS algorithm a false packet rate (FPR) is calculated that has not been done in any previous papers we have reviewed. FPR is the percentage of the ratio of the difference between the packets actually received by the intermediate node and the number of packets forwarded to the actual destination node to the number of packets received by the intermediate node.

Ideally, one needs to confirm that an IDS cannot be compromised. If not, then at least attacks against the IDS must be detected. This is particularly important when using the cluster-based detection approach. If a compromised node happens to be elected as the cluster head, it can launch attacks without being detected because it is the only node that should run IDS, and its IDS may have been disabled already. The works that have been done are based on attacks like packet dropping, malicious flooding, and spoofing [16]. The focus has been on detecting blackhole attacks.

The features that we consider are: (1) route request message from a source node to a cluster head node, (2) route reply message from a cluster head to a source node, (3) routing table updating of each node after a time period, (4) maximum hop count from a cluster head in a cluster, (5) all the packets from a specific source to a specific destination have a distinct sequence number, and (6) a cluster head has all the route information from a source to the destination. The performance graph shows marked improvement as far as packet dropping is concerned. In the future, BHIDS may be extended to the detection of sleep deprivation attacks.

References

1. Debdutta Barman Roy and Rituparna Chaki, MADSN: Mobile Agent Based Detection of Selfish Node in MANET, *International Journal of Wireless and Mobile Networks (IJWMN)*, 3(4), 2011.
2. S. Muthuramalingam, R. RajaRam, Kothai Pethaperumal, and V. Karthiga Devi, A Dynamic Clustering Algorithm for MANETs by Modifying Weighted Clustering Algorithm with Mobility Prediction, *International Journal of Computer and Electrical Engineering*, 2(4), 1793–8163, 2010.
3. S. Chinara and S.K. Rath, TACA: A Topology Adaptive Clustering Algorithm for Mobile Ad Hoc Network, at World Congress in Computer Science Computer Engineering and Applied Computing, 2009.
4. Debdutta Barman Roy, Rituparna Chaki, and Nabendu Chaki, BHIDS: Blackhole Intrusion Detection System in MANET, *Security and Communication Networks*, 2010, pp. 278–288.
5. S. Kurosawa, H. Nakayama, N. Kato, A. Jamalipour, and Y. Nemoto, Detecting Blackhole Attack on AODV-Based Mobile Ad Hoc Networks by Dynamic Learning Method, *International Journal of Network Security*, 5(3), 338–346, 2007.
6. R. Chaki and N. Chaki, IDSX: A Cluster Based Collaborative Intrusion Detection Algorithm for Mobile Ad-Hoc Network, in *Proceedings of the IEEE International Conference on Computer Information Systems and Industrial Management Applications (CISIM 2007)*, 2007, pp. 179–184.
7. Djamel Djenouri and Nadjib Badache, Struggling against Selfishness and Blackhole Attacks in MANETs, *Wireless Communications and Mobile Computing*, 9999, 2007.
8. Kun Xiao, Ji Zheng, Xin Wang, and Xiangyang Xue, *A Novel Peer-to-Peer Intrusion Detection System Using Mobile Agents in MANETs*, IEEE Computer Society, Dalian, China, 2005.
9. Yu-Fang Zhang, Zhong-Yang Xiong, and Xiu-Qiong Wang, Distributed Intrusion Detection Based on Clustering, in *Proceedings of International Conference on Machine Learning and Cybernetics*, 2005, vol. 4, pp. 2379–2383.
10. A. Karygiannis, E. Antonakakis, and A. Apostolopoulos, Detecting Critical Nodes for MANET Intrusion Detection Systems, at 2nd International Workshop on Security, Privacy and Trust in Pervasive and Ubiquitous Computing, 2006.
11. A. Patwardhan, J. Parker, A. Joshi, A. Karygiannis, and M. Iorga, Secure Routing and Intrusion Detection in Ad Hoc Networks, at Third IEEE International Conference on Pervasive Computing and Communications, 2005.
12. Sudarshan Vasudevan, Jim Kuroso, and Don Towsley, Design and Analysis of a Leader Election Algorithm for Mobile Ad-Hoc Network, in *Proceedings of the 19th IEEE International Conference on Network Protocols*, 2004, pp. 350–360.
13. Y.A. Huang, W. Fan, W. Lee, and P.S. Yu, Cross-Feature Analysis for detecTing Ad-Hoc Routing Anoma-lies, in *23rd International Conference on Distributed Computing Systems (ICDCS'03)*, May 2003, pp. 478–487.

14. Y.A. Huang and W. Lee, Attack Analysis and Detection for Ad Hoc Routing Protocols, in *7th International Symposium on Recent Advances in Intrusion Detection (RAID'04)*, French Riviera, September 2004, pp. 125–145.
15. Chunxiao Chigan and Rahul Bandaru, Secure Node Misbehaviors in Mobile Ad-Hoc Network, in *Vehicular Technology Conference*, 2004, vol. 7, pp. 4730–4734.
16. Amitabh Mishra, K. Nadkarni, and Animesh Patcha, Intrusion Detection in Wireless Ad Hoc Networks, *Virginia Tech IEEE Wireless Communications*, February 2004, pp. 48–60.
17. Yi Li and June Wei, Guidelines on Selecting Intrusion Detection Methods in MANET, in *Proceedings of ISECON 2004*, vol. 21 (Newport), 3233 (refereed), 2004 EDSIG.
18. Yi-an Huang and Wenke Lee, A Cooperative Intrusion Detection System for Ad Hoc Networks, in *Proceedings of the ACM Workshop on Security of Ad Hoc and Sensor Networks (SASN'03)*, 2003, pp. 135–147.
19. Richard A. Kemmerer and Giovanni Vigna, Intrusion Detection: An Over View and Brief History, in *IEEE Computer Supplement on Security and Privacy*, 2002, pp. 27–30.
20. Y. Hu, A. Perrig, and D.B. Johnson, Ariadne: A Secure On-Demand Routing Protocol for Ad Hoc Networks, at Proceedings of the Eighth Annual International Conference on Mobile Computing and Networking (MobiCom 2002), September 23–26, 2002.
21. Oleg Kachirski and RatanGuha, Intrusion Detection Using Mobile Agents in Wireless Ad Hoc Networks, in *IEEE Workshop on Knowledge Media Networking (KMN'02)*, 2002, p. 153.
22. M.G. Zapata, *Secure Ad Hoc On-Demand Distance Vector (SAODV) Routing*, IETF Internet Draft (Work in Progress), August 2001, draft-guerrero-manet-saodv-00.txt.
23. S. Basagni, K. Herrin, D. Bruschi, and E. Rosti, Secure Pebblenets, at Proceedings of the 2001 ACM International Symposium on Mobile Ad Hoc Networking and Computing (MobiHoc 2001), Long Beach, CA, October 2001.
24. Mark Slagell, *The Design and Implementation of MAIDS (Mobile Agent Intrusion Detection System)*, Technical Report TR010-07, Department of Computer Science, Iowa State University, Ames, 2001.
25. D. Dasgupta and H. Brian, Mobile Security Agents for Network Traffic Analysis, in *Proceedings of DARPA Information Survivability Conference and Exposition II, DISCEX'01*, 2001, vol. 2, pp. 332–340.
26. Y. Zhang and W. Lee, Intrusion Detection in Wireless Ad-Hoc Networks, in *Proceedings of the 6th Annual International Conference on Mobile Computing and Networking, MobiCom'2000*, pp. 275–283.
27. M.C. Bernardes and E. Santos Moreira, Implementation of an Intrusion Detection System Based on Mobile Agents, in *Proceedings of the International Symposium on Software Engineering for Parallel and Distributed Systems*, 2000, pp. 158–164.

28. G. Helmer, J. Wong, V. Honavar, and L. Miller, *Lightweight Agents for Intrusion Detection*, Technical Report, Department of Computer Science, Iowa State University, 2000.
29. Eugene H. Spafford and Diego Zamboni, Intrusion Detection Using Autonomous Agents, *Computer Networks*, 34(4), 547–570, 2000.
30. C.E. Perkins, E.M.B. Royer, and S.R. Das, *Ad Hoc On-Demand Distance Vector (AODV) Routing*, RFC 3561, July 2003.
31. Dynamic Learning Method, Satoshi Kurosawa, Hidehisa Nakayama, Nei Kato, Abbas Jamalipour, and Yoshiaki Nemoto, Detecting Blackhole Attack on AODV-Based Mobile Ad Hoc Networks, Sendai, Miyagi, Japan (received December 19, 2005; revised and accepted January 27 and March 3, 2006).
32. Sudarshan Vasudevan, Jim Kuroso, and Don Towsley, Design and Analysis of a Leader Election Algorithm for Mobile Ad-Hoc Network, at Proceedings of the 17th IEEE International Conference on Network Protocols (ICNP'04).
33. Latha Tamilselvan and V. Sankaranarayanan, Prevention of Co-operative Blackhole Attack in MANET, *Journal of Networks*, 3(5), 2008.
34. JianYin, Sanjay Kumar Madria, A Hierarchical Secure Routing Protocol against Blackhole Attack in Sensor Network, at Proceedings of the IEEE International Conference on Sensor Network, Ubiquitous and Trust Worthy Computing.
35. J. Zhou, J. Chen, and H. Hu, *SRSN: Secure Routing Based on Sequence Number for MANET*, International Conference on Wireless Communications, Networking and Mobile Computing, WiCom 2007, Shanghai, pp. 1569–1572, 2007.
36. Latha Tamilselvan and V. Sankaranarayanan, Prevention of Blackhole Attack in MANET, at 2nd International Conference on Wireless Broadband and Ultra Wideband Communication.
37. P. Gera, K. Garg, and M. Misra, Opinion Based Trust Evaluation Model in MANETs, *Contemporary Computing Communications in Computer and Information Science*, 168, 301–312, 2011.

7

Intrusion Detection for Wireless Mesh Networks

NOVARUN DEB, MANALI CHAKRABORTY, AND NABENDU CHAKI

Contents

7.1 Introduction

Wireless mesh networks (WMNs) are proliferating as one of the key technologies of the next-generation networks. Security is one of the prime concerns toward actual implementation of any network technology for commercial applications. Network security has intrinsically two approaches: prevention based and detection based. Implementing firewalls or intrusion prevention techniques is often not an attractive solution for energy-constrained network nodes—mobile ad-hoc network (MANET) nodes or mesh clients in wireless mesh networks. However, in the era of pervasive and ubiquitous computing, commercial transactions are performed on the move and over portable devices like cell phones and laptops. These devices have energy constraints, and hence one cannot afford to adopt security measures with high computational overhead. This influences a shift in paradigm from active intrusion prevention to passive intrusion detection. In this chapter, a new cluster-oriented reward-based intrusion detection system (CORIDS) has been described for wireless mesh networks. The performance of the proposed algorithm has been evaluated using the QualNet network simulator. Simulation results also establish superiority of CORIDS over misbehavior detection algorithm (MDA), another recent trust-based intrusion detection system (IDS) for wireless mesh networks, in terms of both higher detection efficiency and lower false positives.

Wireless mesh networks (WMNs) are an extension of existing wireless ad-hoc networks to eliminate the limitations of current network structures and also improve the performance of the overall network. They provide the advantages of both infrastructure-based static networks and infrastructure-less mobile networks. A WMN usually consists of mesh routers and mesh clients. Mesh clients are generally mobile nodes, and they are responsible for the automatic establishment and dynamic upgrading of mesh topology among the nodes. They also act as a router for the other nodes in the network. This makes the network dynamic, scalable, and robust. On the other hand, mesh routers are generally static and provide an infrastructure-based backbone for the WMNs. Mesh routers can integrate different existing wireless networks with the help of gateways and bridges. They also provide network access for both mesh clients and conventional nodes [1].

The backbone of WMNs is formed with the mesh routers. This infrastructure-based backbone can be built using various radio technologies, among which IEEE 802.11 is frequently used. The mesh routers can be connected to the Internet through gateways. They also provide an infrastructure for the client nodes and enable integration of other networks with different radio technologies to the WMN. This approach is called infrastructure meshing. Mesh routers basically use two separate radio frequencies for backbone communication and for client communication.

The client nodes in the WMN provide client meshing. They form a self-configuring and self-healing mesh network among themselves to perform routing of traffic. Client nodes also provide some additional functions, such as self-configuration of nodes and providing end user applications to the customer. The client nodes can communicate with the backbone as well as the other client nodes. Client WMNs are usually formed using one type of radio on devices.

The other advantages of wireless mesh networks include the following. The installation cost is much less than that for the other existing networks. The burden of the system administrator is reduced due to the self-configuring, self-tuning, self-healing, and self-monitoring capabilities of mesh networks. A multihop WMN also eliminates single-point failure and bottlenecks within the network. In advanced mesh networking, a node can go into sleep mode while inactive and then wake up quickly when it becomes active again. This extends the battery life of a node. Increasing scalability is also much simpler and less expensive. The last few years have witnessed WMNs emerging as a leading paradigm for ubiquitous applications. While a large number of the published articles deal with futuristic applications only, some research results are reported toward strengthening the foundations of WMNs. In [2], a novel knowledge plane has been proposed for the WMNs. This knowledge plane is capable of enabling consistent sharing of services ontology among different entities in the WMN. In [3], work has been done on the assignment of frequency bands to radio interfaces toward forming a WMN with minimum interference. A new hybrid, interference, and traffic-aware channel assignment scheme has been proposed in [3] that achieves good multihop path performance between every node and the designated gateway nodes in a multiradio WMN network.

However, the high usage of wireless communication and the presence of a backbone network lead to exposure of WMNs to various abuses and attacks from malicious users and a lack of a clear line of defense. The unreliability of wireless links between nodes and the constantly changing topology lead to increased vulnerabilities in WMNs [4, 5]. This is more critical when the nodes leave and join a network frequently. While various prevention techniques have been designed for securing WMNs, none of them is a silver bullet and holds in a right position with the presence of insiders. Besides, prevention techniques are typically computation-intensive. Commercial transactions are performed over portable devices like cell phones and tablet PCs. Thus, detecting the intrusion and taking appropriate corrective actions emerge as the effective defense line for wireless network security.

Intrusion detection is a second line of defense in network security. Intrusion detection systems (IDSs) have not evolved with the purpose of proactively preventing attacks. Instead, their purpose is to alert network administrators about attacks. An IDS attempts to differentiate the honest and malicious nodes in a network based on the behavior of the nodes [6]. Mitigation of the damage is an expected follow-up action after the IDS successfully detects the intrusion [7]. Detecting intrusions is usually more difficult in the wireless network domain.

7.2 Review of the IDS Solutions for Wireless Networks

A detailed study on the state of the art of intrusion detection systems for mobile ad-hoc networks has been presented in Chapter 5. The chapter also presents a survey of the existing trust-based solutions.

With the birth of wireless mesh networks, however, newer challenges emerged. The hybrid architecture of WMNs is based on fixed infrastructure as well as the client mobility of MANETs. Thus, researchers tried to adopt or extend the existing intrusion detection algorithms for MANETs. A misbehavior-based intrusion detection scheme for WMNs was proposed in [13]. The algorithm reduces false positives by selecting an appropriate value of threshold, but cannot eliminate it. The number of false positives increases with an increase in the number of malicious clients within a fixed threshold value. RADAR [14] was proposed as a reputation-based intrusion detection scheme. Routing loop attacks are detected with high false positives. RFIDS [15] uses one or more RF transmitters that emit radio frequency (RF) into space, along with a well-planned network of RF receivers for detection. The proposed method gives better clarity, higher frequency, and higher speed of processing. A lightweight PCA-based intrusion detection system was proposed in [16], such that appropriate selection of threshold values can reduce the number of false alarms considerably. However, the authors do not mention anything about network traffic or false negatives. Another lightweight intrusion detection system was suggested in OpenLIDS [17]. The solution also avoids false positives. However, in the process, it fails to detect User Datagram Protocol (UDP) denial-of-service (DoS) flood attacks. Another two IDSs were proposed for defending selective forwarding attacks [18] in WMNs and detecting selfish nodes in WMNs [19]. They both claimed to have a high detection rate while having a low rate of false positives, but there will be always some detection inaccuracy due to some unrealistic assumptions. A community intrusion detection and prewarning system based on wireless mesh network was presented in [20]. The system can establish a network automatically by using the advantage of multihop communication of mesh networks. Redundant nodes can be placed in important areas to ensure reliability and resist destruction. Another community-based IDS [21] presents a set of sociotechnical challenges associated with developing an intrusion detection system for a community wireless mesh network. But this architecture does not explicitly address the challenges with the system administrator. Another IDS was proposed based on Kohonen networks [22]. Even if this solution is quite effective in terms of the number of false positives, it involves high computation overhead.

Both the referred works, [3] and [10], are inefficient in terms of energy consumption. CORIDS is a very attractive solution in terms of energy efficiency. This is where our algorithm is so lucrative. We relieve the mesh clients from participating in the intrusion detection process. Only the backbone routers/cluster heads participate in the intrusion detection process, making our algorithm extremely lightweight. [11] is also not efficient in terms of energy. Also, it suffers from false negatives in direct proportion to the number of malicious nodes. CORIDS is extremely lightweight, and the percentage of false negatives varies slightly with node density and node mobility, but hardly with the percentage of malicious nodes in the network. SCAN [7] suffers from both high false positives and high false negatives when the nodes have high or low mobility, respectively. CORIDS has a low percentage of both false positives and false negatives with varying node density and node mobility. HIDS [4] reduces the communication overhead compared to SCAN, but again, it is not lightweight, as every node has to maintain a table and constantly update that table. CORIDS performs much better than HIDS in terms of energy efficiency, as the energy-constrained mesh clients have to maintain only a pair of counters and regularly update them as and when packets are sent or received.

CORIDS also performs significantly well in comparison to the existing intrusion detection systems for wireless mesh networks. Both MDA [9] and RADAR [5] report high false positives in direct proportion to the percentage of malicious clients, whereas CORIDS is stable in the number of false positives, irrespective of change in the percentage of malicious nodes. The PCA-based intrusion detection system proposed in [8] does not mention anything about false negatives. CORIDS gives a detailed analysis of the false positive percentage. OpenLIDS [6] has a major drawback in that it fails to detect UDP DoS attack. On the other hand, CORIDS has been well tested and successfully detects all types of DoS attacks.

After making this comprehensive survey on existing intrusion detection systems for MANETs and WMNs, we observe that cluster-based IDSX [8] and trust-based collaborative BHIDS [12] for MANET offer several advantages over most of the other IDSs mentioned above. However, even if some trust-based IDSs are already proposed, the cluster-based architecture is unexploited for WMNs. Besides, both IDSX and BHIDS have their limitations, as described above. Neither of the approaches can be readily used for efficient intrusion detection in WMNs. In this chapter, some of the basic ideas have been taken from both IDSX [8] and BHIDS [12] to propose a new cluster-oriented reward-based IDS (CORIDS) algorithm for WMNs. The proposed algorithm avoids the drawbacks of IDSX and BHIDS by elegant utilization of WMN backbone. In the proposed CORIDS solution, the mesh clients are relieved from doing any kind of trust-based computations. Only the mesh routers have to store the tables for their corresponding mesh clients and do the necessary computations.

7.3 Introduction to CORIDS

The following sections explain in detail the cluster-based architecture that has been proposed for CORIDS, how the mesh routers and mesh clients behave in this structure, what are the different types of attacks that are being addressed, and finally, how the CORIDS algorithm works in detail.

7.3.1 Cluster Architecture for WMNs

The hybrid architecture of WMNs has been assumed. Mesh clients have mobility and resource constraints. Mesh routers are static and have resources that can be replenished. The mesh clients are organized in a hierarchical model. The entire network is divided into several disjoint or overlapping clusters. The clients are divided into clusters,

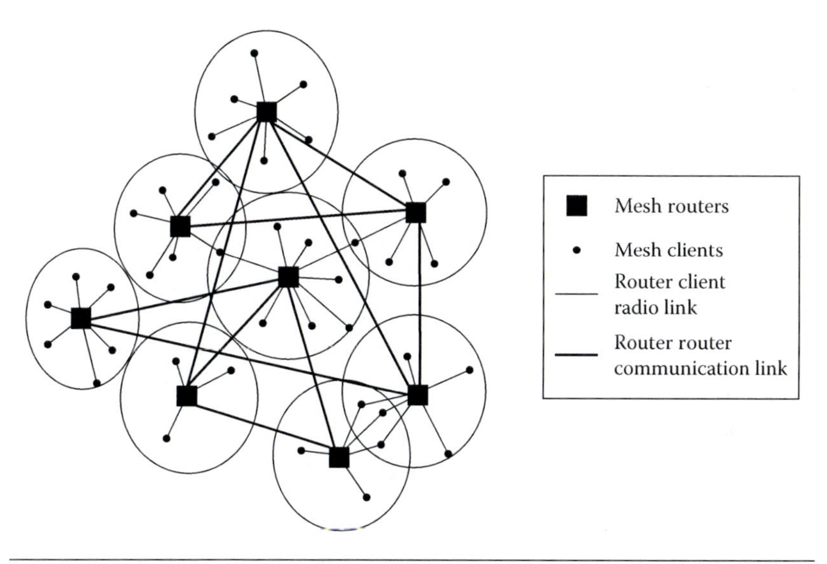

Figure 7.1 The proposed cluster-based wireless mesh network. (From N. Deb, M. Chakraborty, and N. Chaki, 2013. *Security and Communication Network*, John Wiley & Sons. With permission.)

with each cluster having one mesh router as the cluster head (CH). Clusters are sufficiently spaced out so that no cluster in the network has more than one CH. A cluster is defined by the CH and all mesh clients that are within one-hop distance (in the radio frequency range) of the CH. The total number of clusters is bounded by the number of mesh routers in the WMN backbone. Client members of a cluster interact directly with the CH since they are in the direct radio range of the CH. Membership information about a cluster is maintained by the CH and is updated at regular intervals (Figure 7.1).

Cluster members send packets to other cluster members of the same or different cluster. These packets reach the CH of the cluster where the source resides. The CH checks its membership information with the receiver's address. If the receiver is in the same cluster, the CH sends it to the destination node. If the destination node belongs to some other cluster, then the CH checks its routing table information to decide where to route the packet. Intercluster communication is achieved when CHs share their information. Redundant multihop paths exist between CHs.

Only a CH can participate in the intrusion detection process. Each CH monitors the behavior of its cluster members. CHs store additional information about suspected nodes. Also, CHs store trust information of its cluster members. Since mesh clients have limited memory and power resources, we do not involve the cluster members in the intrusion detection process. Only CHs will run the CORIDS algorithm. CHs exchange trust information of their members among themselves at regular time intervals. Each CH updates the trust values of its members.

CHs can declare a node as malicious based on some parameters and their threshold values. A member can be declared as malicious by any CH through which the cluster member has transmitted or received packets. It must be kept in mind that mesh clients have mobility. As a result, a mesh client may send packets of some application through several CHs. Thus, behavior-related information about a mesh client may get distributed over several CHs. If a node is malicious, it may misbehave intelligently enough such that none of the CHs cross the threshold value for some parameter. The malicious mesh client may remain undetected. However, by combining information from all CHs, the misbehaving node can be detected as malicious.

Since members of the mesh backbone are the CHs, they are static and have renewable resources, great computational capability, and no power constraints. These properties of mesh routers have been fully utilized to relieve the mesh clients from doing any kind of computations in the intrusion detection process. Very little resource utilization of mesh clients is performed, as will be seen later. The various jobs that a CH is expected to do include storage of information, routing, identification of the cluster boundary, and distinguishing malicious nodes from good ones.

CHs maintain two different parameters for each of its cluster members: the packet arrival rate (PAR) and the packet delivery rate (PDR). Since each packet originating from a source client is sent to the corresponding CH, and each packet is delivered by a CH to the destination client, PAR and PDR values for each client are maintained at the corresponding CH. Cluster members also keep track of the number of packets sent and received by it. This is done with the help of two other parameters: packets sent (PS) and packets received (PR). Both these counters are maintained in the mesh clients by kernel level protocols. It is assumed that an attacker cannot tamper with these values, as the packets are encrypted using some standard encryption schemes. Ordinary member nodes send their PR and PS information to their current CHs at regular time intervals. Whenever the PAR, PDR, PR, or PS values of a mesh client are not within their respective thresholds, an intrusion is suspected. Trust values are updated based on these values, and if the trust value for a mesh client falls below a threshold value, then it is declared as a malicious node.

During routing, when a CH receives a packet from one of its members, it updates the PAR parameter associated with that member. When the CHs deliver packets to their cluster members, they update the PDR parameter associated with the respective destination node. Original CHs may have to forward a packet to another CH based on its routing table information. An intermediate CH, on receiving a packet from another CH, checks its routing table information to decide whether to forward the packet or to deliver it to one of its members. The underlying routing protocol that is assumed for CORIDS is the Ad-Hoc On-Demand Distance Vector (AODV).

7.3.2 The CORIDS Algorithm

To relieve the mesh clients from consuming their resources on intrusion detection, the CORIDS algorithm is executed only in the mesh routers. The parameters involved in the intrusion detection process are defined as follows.

- TR_{val}: Trust value of a cluster member as evaluated by its CH.
- TH_{Tr}: Threshold trust value; if the trust value of a mesh client falls below TH_{Tr}, it is declared malicious.
- PAR_X: Packet arrival rate for cluster member X as maintained by its CH.
- PDR_X: Packet delivery rate for cluster member X as maintained by its CH.
- TH_{PAR}: Threshold value of packet arrival rate.
- N_X: Unique ID of node X.
- PS: Number of packets sent to the CH.
- PR: Number of packets received from the CH.

Before starting the working of the algorithm, it is mandatory to mention the assumptions behind the proposed algorithm:

1. Every CH overhears the activity of its cluster members.
2. All mesh clients and mesh routers have a global unique ID.
3. CHs are relatively secured, or they have enough resources to implement different layers of security.
4. The threshold values are precalculated and set for the entire network. These values are stored in the CHs. Depending on the priority of clusters, different CHs may have different values set as their threshold.
5. The PR and PS counters at the client side are maintained by the operating system, and these values cannot be tampered with.

When a client enters the network, it must wait for itself to become a part of a network cluster. As soon as it receives a HELLO packet from some CH, it immediately responds with an ECHO packet. It is only then that it becomes a part of that cluster, and thus a part of the network. Whenever a CH identifies new members in its cluster boundary, it assigns them a globally unique ID and an initial value of TR_{val}. The trust value of each cluster member is stored in the trust table of the CH. Monitoring of the network is performed every time a packet is received by the CH from a member or delivered by the CH to a member. All data and control packets are signed by the source of the packet. Whenever a data packet reaches a CH from a cluster member X, it only attaches the unique ID of the client with that packet. This packet is then propagated through the network.

The CORIDS algorithm provides a three-step solution to the intrusion detection problem. In the first phase of the algorithm, CHs collect information about their cluster members, which is stored in the CHs of other clusters. This phase is necessary, as it may so happen that a cluster member, due to its mobility, sends data packets through several CHs. If the node is under the influence of a DoS attack, it may try to block network resources by generating excessive traffic. It is a basic assumption while defending any kind of attack that the attacker is very intelligent. So the malicious node may start sending too many packets but evenly route them through different CHs. Thus, if a cluster member has multiple CHs, then its PAR information will most likely be distributed over these CHs. Aggregating this information before making any kind of decision is mandatory.

Merging of information from different cluster heads becomes necessary when a cluster member receives or sends packets through different cluster heads. PAR/PDR information for a cluster member, as stored in a cluster head, reflects only the number of packets received or delivered to that cluster member via that cluster head. However, PR/PS information stored in a cluster member represents the history of packets received by and sent from the node. So in order to correctly assess the behavior of cluster members, we need to merge the PAR/PDR information of mesh clients stored in the CHs.

At the beginning of every epoch, CHs send IDS_Request packets to all their cluster members on one frequency and IDS_Update packets to all neighboring cluster heads on another frequency. Each cluster head stores PAR/PDR information in a table, along with the particular cluster member's node address. IDS_Update packets contain a list of node addresses along with their PAR/PDR information as collected by that cluster head. So when a cluster head receives an IDS_Update packet from another cluster head, it scans through this list and checks if there is any node that is a member of both the clusters as represented by these two cluster heads. All distributed information about a particular cluster member (if that member belongs to more than one cluster) is merged in all the cluster heads that have information about that cluster member.

In the second phase of the algorithm, a CH interacts and collects control information from its cluster members. After collecting information from its cluster members, the CH executes the CORIDS algorithm in the third phase. The CHs update the trust values associated with each of their cluster members. Once trust values are updated, cluster members can be declared malicious if required. Cluster members that are declared as malicious are reported to other CHs. The following three algorithms depict phases 1, 2, and 3 of the CORIDS algorithm.

Algorithm: Phase 1

1.1. CHs store PAR, PDR, PR, and PS information of cluster members with their corresponding IDs.
1.2. After a fixed time slice, CHs broadcast their respective client information to all other CHs in the network.

1.3. When a CH receives such an update packet, it scans the packet and checks to see if information is available about some node that also belonged to its own cluster at some point in time. This is done by checking the node ID N_x associated with each parameter set.
1.4. If any N_x matches with those of its cluster members, then it updates the information associated with that node ID, N_x.
1.5. Steps 1.3 and 1.4 are performed by all CHs.
1.6. Proceed to phase 2.

Algorithm: Phase 2

2.1. Mesh clients within every cluster send the values of their PS and PR parameters to their respective CHs.
2.2. Mesh clients send their PR and PS data along with their respective node ID, N_x.
2.3. The CH maintains a table of all cluster members within its own cluster.
2.4. Proceed to phase 3.

Algorithm: Phase 3

3.1. CHs compare the PAR and PDR values of a cluster member with its corresponding PR and PS values, and also their respective thresholds.
3.2. Attacks are classified based on these values.
3.3. Rewards are calculated for all nodes. Normally behaving nodes are positively rewarded, whereas misbehaving nodes are negatively rewarded.
3.4. Trust values of cluster members are updated using the evaluated reward values.
3.5. The updated trust values are compared with a predefined threshold trust value. If the trust value of any cluster member falls below the threshold value, then it is identified as a malicious node.
3.6. Stop.

The algorithm is executed by the CHs in a distributed manner and at random but regular time intervals.

7.3.3 Handling Attacks Using CORIDS

The basic motivation behind the algorithm is that trust values of nodes are updated based on current information (reward based). The trust values of compromised nodes should be decreased, while those of normal nodes should be increased. When a node enters the network, it is assigned an initial trust value given by Equation 7.1:

$$TR_{VAL} = (2^n - 1)/2 \tag{7.1}$$

Here n is the number of bits assigned for storing trust values. A node whose trust value falls below the threshold value of trust (TH_{Tr}) is declared malicious. Let us look into some of the different situations and attacks that may occur, how the attacks are detected, and how clients are rewarded. In the simulation, the trust value is initialized to 16,384 for all the client nodes. An integer variable has been used to represent trust. The values for a 2-byte integer range from –32,768 to 32,767. We assign the midway value in the positive scale, i.e., 16,384. It is equivalent to 0.5 on a normalized scale of [0, 1].

The packet transfer rate is taken as one per second. Since our IDS algorithm repeats every 20 s, ideally a node should generate a maximum of 20 packets in between successive epochs. A buffer of five more packets is allowed, keeping in mind that traffic can be heavy at times. If the number of packets is more than 25, we decrease the trust value of a node by 300*(number of extra packets generated). This value of 300 has been decided arbitrarily. When trust value falls below the threshold value of 10,000, we declare a node as malicious. The threshold value is decided such that we can detect an attacker as soon as possible. This is how denial-of-service attacks are detected. If a node behaves normally, then its trust value increases by a fixed amount, as defined in Equation 7.3. The blackhole attacks are detected following a similar approach:

A node behaves normally: Good nodes are characterized by the property that their PAR, PDR, PS, and PR values are consistent. Good nodes should be duly rewarded. A node X is assumed to behave normally if $PAR_X < TH_{PAR}$, $PAR_X = PS$, and $PDR_X = PR$. Such a node is rewarded as defined in Equation 7.2:

$$Reward = R \times (PAR_X + PDR_X)/2 \tag{7.2}$$

DoS attack (generation of spurious packets): A denial-of-service attack is primarily a resource utilization attack. It causes congestion in the available network resources. One such type of DoS attack is spurious packet generation by flooding packets throughout the network. The DoS attack can be distributed in nature if several nodes start generating spurious packets from different points of the network. A node X is definitely under DoS attack if the packet arrival rate for that node exceeds the threshold value, i.e., $PAR_X > TH_{PAR}$.

$$\text{Reward} = -R \times (PAR_X - TH_{PAR}) \tag{7.3}$$

Blackhole attack (routing misbehavior by dropping packets): The routing protocol has been so modified that packets are routed only through CHs. Since CHs are intrinsically secure, a blackhole attack can be launched only during route setup. The attacker can maliciously claim to be the final destination for an application and generate a route reply packet. Once a route is set up between the sender and the attacker, all packets would be sent to that attacker node.

However, CHs increment their PDR values only when the destination address matches with that of its cluster member. Thus, although the packet is delivered to the malicious node, its PDR value is not incremented at the CH. On the other hand, whenever a cluster member sends or receives a packet, its PS or PR parameter gets incremented. Thus, there is a mismatch between the PR and PDR values of the malicious node at the CH. A node X is surely a blackhole attacker if the packet delivery rate for the node is less than the packet received parameter, i.e., $PDR_X < PR_X$.

$$\text{Reward} = -R \times (PR_X - PDR_X) \tag{7.4}$$

Using the different parameter values, the rewards that every cluster member has acquired since the last execution of the algorithm can be evaluated. Once the rewards for all cluster members are obtained, their trust values can be recomputed and reevaluated. The trust value of a cluster member is updated according to Equation 7.5.

$$TR_{VAL\_x}(t) = TR_{VAL\_x}(t-1) + Reward_X \quad (7.5)$$

After trust updating, it is checked if $TR_{VAL\_x}(t) < TH_{Tr.}$ Based on this result, a node is declared either malicious or safe.

7.4 Simulation Results and Performance Analysis

The findings are based on simulations of a WMN network model in Qualnet. The simulation scenario and settings are listed in Table 7.1.

Evaluation metrics: A common criterion for evaluating an anomaly detection scheme is the trade-off between its capability of detecting anomalies and the ability of suppressing false alerts. In the experiment, the detection efficiency and false positive percentage of CORIDS are examined. Ten simulation epochs were executed for averaged results. A trend of the average number of iterations required for each experiment is also shown. Also, the results have been compared with the MDA algorithm [13] in terms of false positive percentage and detection efficiency.

Evaluation of performance parameters: Some of the nodes have been arbitrarily set as malicious during simulation. The percentage of malicious nodes varies from 10 to 50%. For detecting both blackhole and DoS attacks, one set of readings has been taken by varying node density, and another set of readings by varying mobility. Now, for example, when node density is 50, 5 sets of data are taken by increasing the number

Table 7.1 Simulator Parameter Settings

PARAMETER	VALUE
Terrain area	1500×1500 m^2
Simulation time	200 s
MAC layer protocol	Distributed coordination function (DCF) of IEEE 802.11b standard
Network layer protocol	AODV routing protocol
Traffic model	CBR
Number of CBR applications	10% of the number of mesh clients
Mesh router:mesh client	1:5
Types of attacks implemented	DDoS attack using spurious packet generation and routing misbehavior attack using blackhole
Mobility model	Random waypoint
Initial trust value of nodes	16,384

of malicious nodes from 5 (10%) to 10 (20%), 15 (30%), 20 (40%), and 25 (50%). During every simulation, the number of malicious nodes is known, and that many nodes are arbitrarily set as malicious from the configuration file. Since which nodes are supposed to behave maliciously are known, this can be checked with the intrusion data generated by CORIDS.

Since CORIDS is a reward-based IDS, either positive or negative rewards are assigned to the nodes. Depending on the rewards earned, the algorithm adjusts the trust values of the nodes. When trust values fall below a threshold, the corresponding nodes are declared as malicious. This is what comprises the intrusion data of our algorithm. So at the end of the experiment, we check to see which nodes have been detected as malicious by CORIDS.

Suppose x nodes had been set as malicious before simulation, and CORIDS detects y nodes as malicious at the end of the experiment, out of which z nodes belong to the list of malicious nodes set during configuration ($z \leq y$, $z \leq x$). Then the detection efficiency can be calculated as

$$\text{Detection efficiency} = z/x * 100 \tag{7.6}$$

If $z = x$, then detection efficiency is 100%. Now, $(y - z)$ may be greater than or equal to 0. If y is greater than z, then this implies that CORIDS has detected some benign nodes as malicious. These are the false positives, which can be expressed as a percentage as follows:

$$\text{False positive} = (y - z)/x * 100 \tag{7.7}$$

If $y = z$, then there are no false positives. This is how detection efficiency and false positives are calculated.

A false negative is closely related to detection efficiency. Detection efficiency is a measure of the number of malicious nodes correctly identified as intrusions, and a false negative is a measure of the number of malicious nodes that could not be detected by the IDS. Since detection efficiency is expressed as a percentage, false negative can be easily calculated as

$$\text{False negative} = (100 - \text{Detection efficiency}) \tag{7.8}$$

In the simulation settings, constant bit rate (CBR) traffic is generated at the rate of one packet per second, and the epoch duration is 10 s. So in every epoch, 10 packets are in transit—either sent, received, or forwarded. So we consider a tolerance of 30%, which is three packets. A node may drop up to three packets or generate three extra packets in one epoch. This is classified as normal behavior, wherein the node has its trust increased. If a node drops or generates four or more packets more than normal, then it is classified as malicious behavior and its trust is decreased proportionally to the number of extra packets dropped or generated.

Results and analysis: The CORIDS algorithm is simulated in several stages. First, blackhole attacks are implemented and the performance of CORIDS is tested. The performance metrics have already been mentioned above. In the next stage, CORIDS is simulated for detecting distributed denial-of-service (DDoS) attacks. The same set of readings is taken. Finally, the performance of CORIDS is compared with the existing MDA [13] algorithm.

Blackhole attack: First, CORIDS is tested by implementing the blackhole attack. This attack has been implemented in the simulated environment as follows. When there exists a CBR application, the first step is for the sender to discover a path to the receiver. When the path is being set up, if a malicious node receives a RREQ_packet, then it immediately responds by returning a RREP packet, claiming to be the receiver. As a result, the path is set up from the sender to the blackhole.

A critical analysis has been done for all the graphs. First, the performance of CORIDS for blackhole attack detection is measured with variation in mobility. Figures 7.2a and b show the results. The false positive percentage remains in the range of 0–5% (Figure 7.2a). The detection efficiency lies in the range of 85–100% (Figure 7.2b). Using the above formula (7.8), the false negative percentage comes out to be in the range of 0–15%. Each of these parameters has been evaluated by varying mobility from 10 to 90 mps, and varying the percentage of malicious nodes for each node mobility value. All

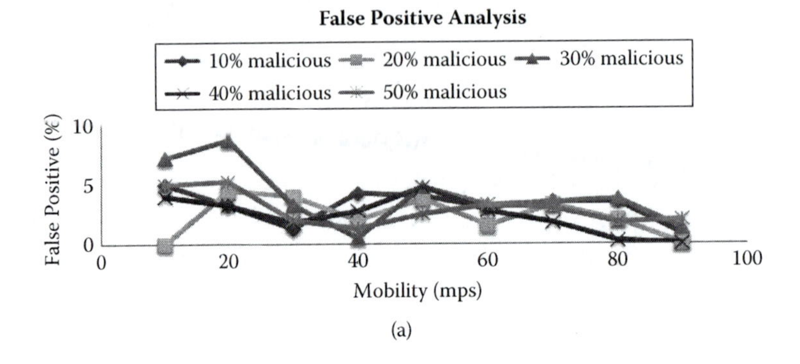

(a)

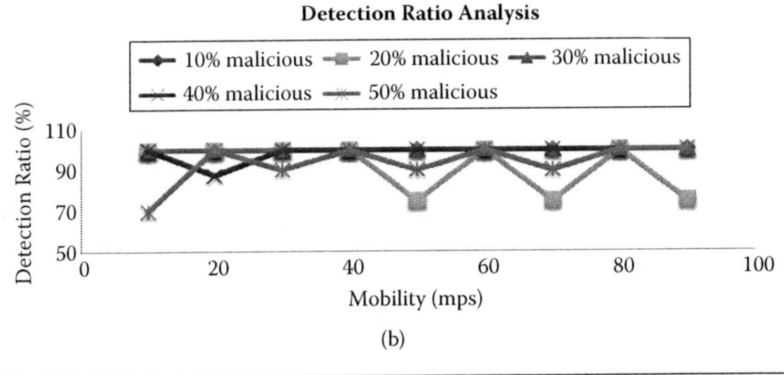

(b)

Figure 7.2 (a) False positive analysis for blackhole attacks with variation in node mobility. (b) Detection efficiency analysis for blackhole attacks with variation in node mobility. (From N. Deb, M. Chakraborty, and N. Chaki, 2013. *Security and Communication Network*, John Wiley & Sons. With permission.)

these parameters have reasonably good values for CORIDS to qualify as a good intrusion detection algorithm.

A different set of data is then taken for detecting blackhole attacks. This time the mobility of the nodes is kept constant at 30 mps and the density of nodes is varied from 10 to 45 in steps of 5. For each data set, readings are taken by changing the density of malicious nodes from 10 to 50%. This time more consistent behavior is observed. Figures 7.3a and b show the results. The false positive percentage remains in the range of 0–4% (Figure 7.3a). The detection efficiency is seen to lie in the range of 85–100% (Figure 7.3b). Using the above formula (7.8), it is seen that the false negative percentage comes in the range of 0–15%. Each of these parameters has been evaluated by varying node density and varying the percentage of malicious nodes for each node density.

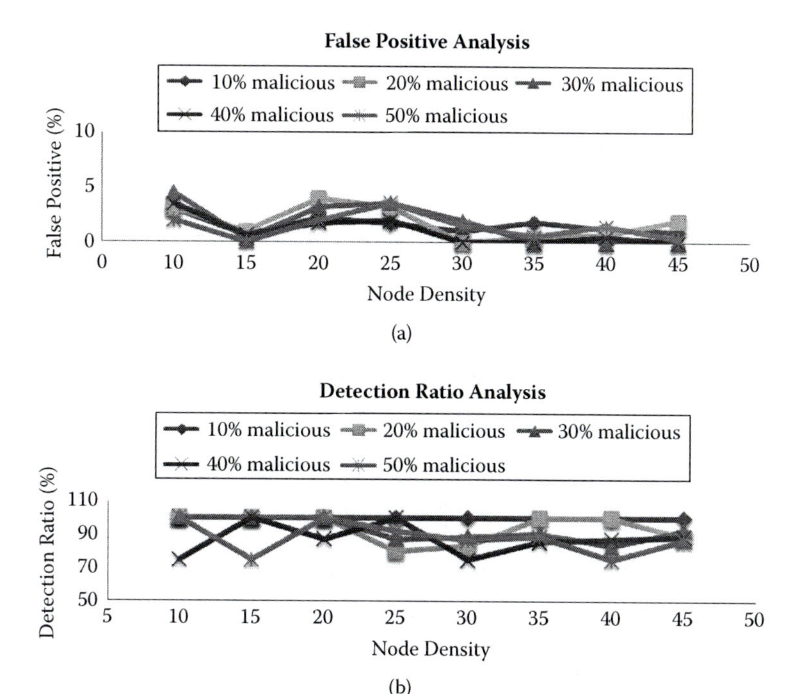

Figure 7.3 (a) False positive analysis for blackhole attacks with variation in node density. (b) Detection efficiency analysis for blackhole attacks with variation in node density. (From N. Deb, M. Chakraborty, and N. Chaki, 2013. *Security and Communication Network*, John Wiley & Sons. With permission.)

All these parameters have reasonably good values for CORIDS to qualify as a good intrusion detection algorithm.

It is seen that the number of epochs required by CORIDS for malicious node identification is almost constant over varying mobility. When the performance of CORIDS is analyzed in terms of the number of epochs required, some sort of stability is observed. Figures 7.4a and b show the graphs. While detecting blackhole attacks, varying mobility keeps the average number of epochs in the range of 4.5 to 6.5 (Figure 7.4a). Varying the node density keeps the average number of iterations in a range between 5 and 7 (Figure 7.4b). This is a significant advantage, as mobility usually affects the performance of most intrusion detection algorithms. The number of epochs of CORIDS required as the node density increases is slightly increasing. However, the rise is not sharp. Thus, there

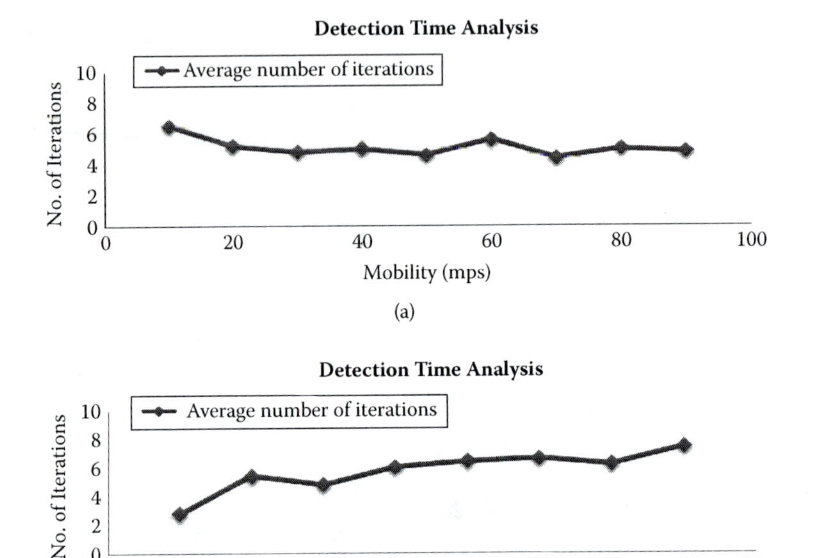

Figure 7.4 (a) Detection time analysis for blackhole attacks with variation in node mobility. (b) Detection time analysis for blackhole attacks with variation in node density. (From N. Deb, M. Chakraborty, and N. Chaki, 2013. *Security and Communication Network*, John Wiley & Sons. With permission.)

is not much variation in the number of iterations required for detecting malicious nodes.

There are two scenarios in which a node behaves normally to increase its trust value and then starts misbehaving. In the first case, the node increases its trust value as soon as it joins the network, and then starts behaving maliciously consistently. In such a situation, the only way that this affects CORIDS is in the detection time. It will take a greater number of iterations/epochs to reduce the increased trust value of the malicious node below the threshold.

A more serious situation is the selective forwarding attack or grayhole attack where a node drops packets selectively. The node is positively rewarded for the majority of packets that it forwards toward the destination and negatively rewarded for the packets that it drops. As a result, the trust value of the node will never fall below the threshold if the attacker intelligently manages the ratio of forwarded packets to dropped packets. In such a situation, CORIDS fails to detect the attacker, and it is counted as a false negative.

Distributed denial-of-service attacks: DoS attacks can be implemented in several ways. In the simulation, one is implemented as the generation of spurious packets by a malicious node, thereby leading to congestion. Network services can no longer be provided to mesh clients. Thus, spurious packet generation can lead to denial of service. The attack is implemented in a distributed manner in the sense that several nodes in the network can start generating spurious packets simultaneously.

The false positive analysis for the CORIDS algorithm is very impressive in the case of distributed denial-of-service (DDoS) attack. Change in mobility does not affect the performance of the algorithm. This is depicted by the linear nature of the curves shown in Figure 7.5a. Also, the close

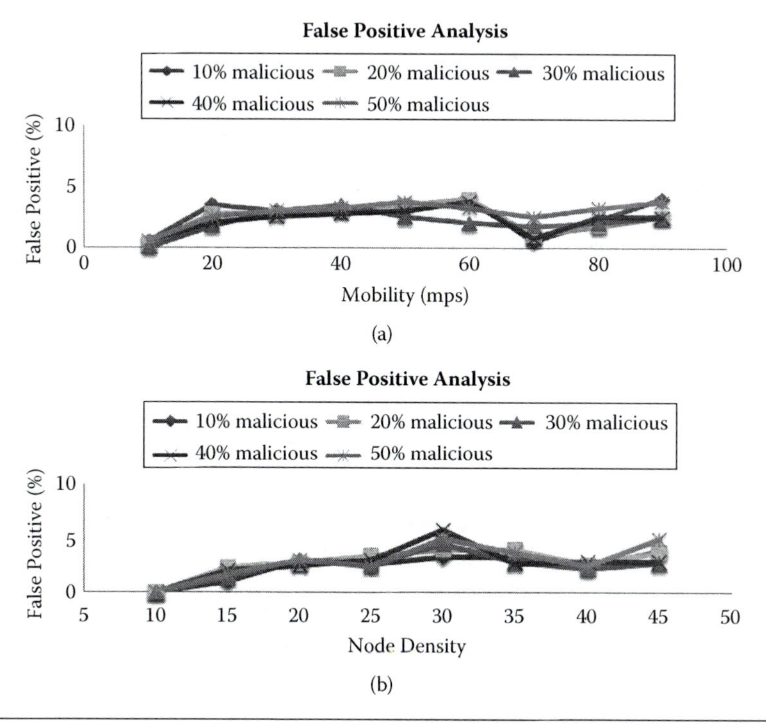

Figure 7.5 (a) False positive analysis for DoS attacks with variation in node mobility. (b) False positive analysis for DoS attacks with variation in node density. (From N. Dob, M. Chakraborty, and N. Chaki, 2013. *Security and Communication Network*, John Wiley & Sons. With permission.)

proximity of the five different lines brings out the consistent behavior of the algorithm in spite of varying the density of malicious nodes. The algorithm also produces good results with varying node density and keeping the mobility constant at 30 mps. The false positive percentages are restricted in the very low and narrow range of 0–4%. Figure 7.5b plots the results as the graph. The behavior of the algorithm is very consistent in spite of increasing node density and percentage of malicious nodes.

Unlike blackhole attacks, the detection efficiency of CORIDS is not plotted for DDoS attacks as *the detection efficiency is always 100%.* This is a significant result that needs to be highlighted. The number of epochs required for identifying DDoS attackers is again a linear function when plotted against variation in mobility. While detecting DDoS attacks, varying mobility keeps the average number of epochs in the range of 3 to 5 (Figure 7.6a). Varying the node density keeps the average number of iterations in the range between 3.5 and

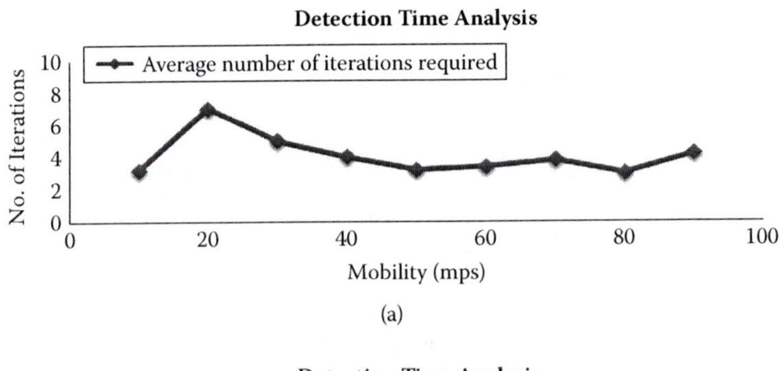

(a)

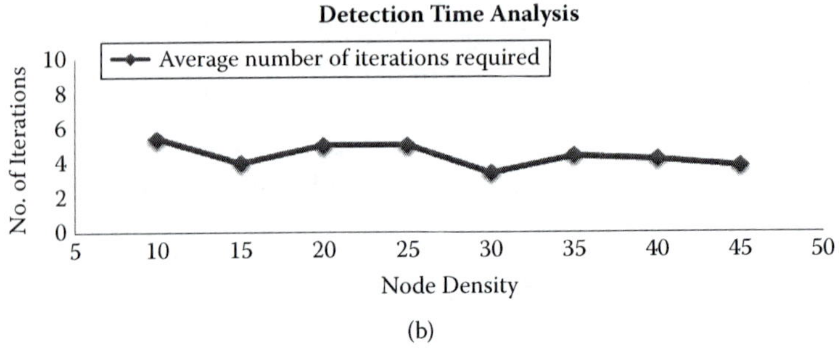

(b)

Figure 7.6 (a) Detection time analysis for DoS attacks with variation in node mobility. (b) Detection time analysis for DoS attacks with variation in node density. (From N. Deb, M. Chakraborty, and N. Chaki, 2013. *Security and Communication Network*, John Wiley & Sons. With permission.)

5.5 (Figure 7.6b). This reiterates the fact that an increase in node mobility does not degrade the performance of the algorithm significantly. A huge plus point. The number of epochs for detecting DDoS attacks in the presence of varying node density is also almost constant. This adds to the effectiveness of the CORIDS algorithm, as most intrusion detection systems falter under heavy loads.

Performance variation with different threshold values: Figure 7.7a shows the variation of detection efficiency with change in threshold values. The performance of CORIDS is tested with both higher and lower values of threshold. The higher the threshold value, the more sensitive the system is to attacks. So

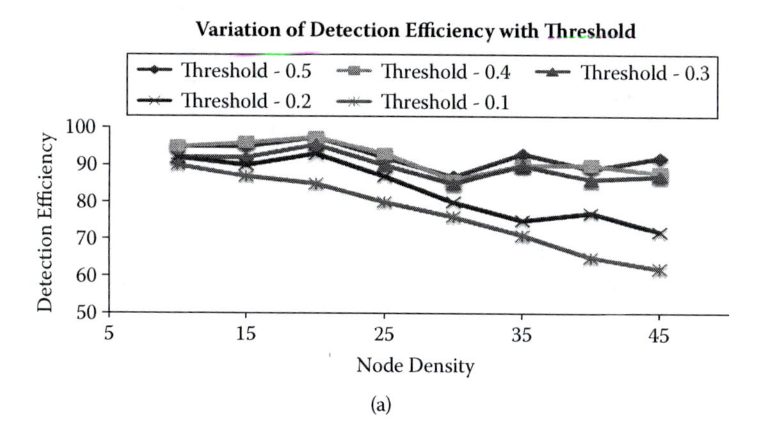

(a)

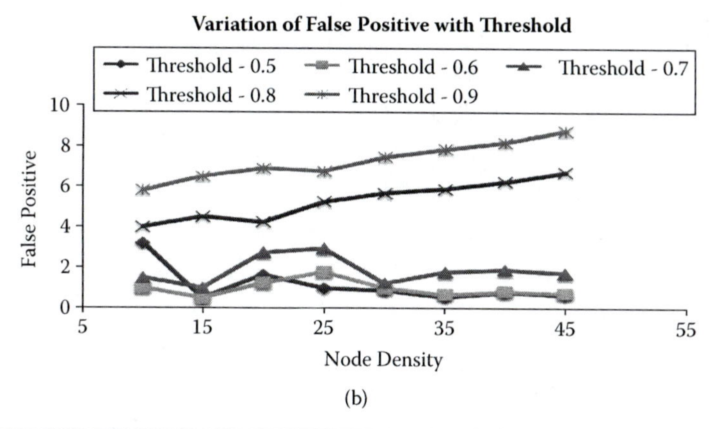

(b)

Figure 7.7 (a) Variation of detection efficiency with variation in node desnity for different thresholds. (b) Variation of false positive with variation in node density for different thresholds. (From N. Deb, M. Chakraborty, and N. Chaki, 2013. *Security and Communication Network*, John Wiley & Sons. With permission.)

detection efficiency is not affected by higher threshold values. This is clear from the graph as well. For threshold values of 0.3 and above, the detection efficiency of the system remains above 85%. But for lower threshold values, the detection efficiency starts decreasing, as it requires a sufficient amount of misbehavior from the malicious nodes to decrease their trust values below the threshold. Thus, it can be concluded that for threshold values below 0.3, the detection efficiency decreases or the false negative increases.

The performance of the CORIDS algorithm is also tested against larger threshold values, and it has been observed that it affects the false positive percentage. Small variations in PAR-PS statistics or PDR-PR statistics were sufficient to decrease the trust value of the nodes below the threshold and declare them as malicious. This is visible in the second graph (Figure 7.7b) that is obtained from simulation.

The graphs in Figures 7.7a and b show that detection efficiency remains considerably low for threshold values of 0.3 and below. When the threshold is set above 0.7, the false positive percentage increases, and also increases with increase in node density. Thus, these simulation results show that the performance of CORIDS remains stable and efficient in the threshold range of 0.3 to 0.7. For values of threshold below this range, the detection efficiency decreases, and for values of threshold above this range, the false positive percentage increases.

Comparison with MDA [13] algorithm: Finally, these data are taken and compared with the performance of the MDA [13] algorithm published in 2008. The MDA algorithm was chosen because, like CORIDS, it is also a trust-based intrusion detection system for WMNs. The MDA algorithm declares a node as malicious whenever its trust value falls below a certain threshold, similar to the CORIDS algorithm. We compared the performance of the two algorithms under varying densities of malicious nodes. The evaluation metrics were the same—false positive percentage and detection efficiency.

The working of CORIDS has been explained in detail in Section 7.3. Let us look into the working principle of MDA. MDA assumes that any two mesh clients x and y communicate through a common set of routers. This common set is defined by those sets of routers that have previously communicated messages between x and y *or* have separately communicated messages from x and y. In other words, this common set refers to those routers that have an existing trust history of the clients x and y. These are the routers that participate in the misbehavior detection algorithm.

The past trust values T_x and T_y are calculated for the mesh clients x and y. MDA divides the common set of routers {M} and trust values T_x and T_y into g groups ($g \geq 1$) as $\{\{T_{x1}\}, \{T_{x2}\}, \ldots, \{T_{xg}\}\}$and $\{\{T_{y1}\}, \{T_{y2}\}, \ldots, \{T_{yg}\}\}$. Here g represents the number of routers in the common set {M}. T_{xk} represents the trust value of mesh client x as evaluated by mesh router k. The trust values are then arranged according to groups as $\{\{T_{x1},T_{y1}\}, \{T_{x2},T_{y2}\}, \ldots, \{T_{xg},T_{yg}\}$, and the correlation is calculated using the following equation:

$$\rho(Tx,Ty) = cov(Tx,Ty)/\sigma_{Tx}\, \sigma_{Ty} \qquad (7.9)$$

where σ_{Tx} and σ_{Ty} are the standard deviations of clients x and y. MDA then calculates the average correlation:

$$\rho_{avg} = \Sigma_{i=1 \text{ to } g}\, (\rho_i/g) \qquad (7.10)$$

Finally, the average correlation thus calculated is compared with a predefined threshold value. If $\rho_{avg} \leq$ threshold, the mesh client y is declared as malicious.

A detailed analysis reveals why it was decided to compare CORIDS with MDA. First, both CORIDS and MDA are based on threshold-based trust evaluation. That is, in either case, a node is declared as malicious if its trust value falls below a certain threshold. This implies that both the algorithms are vulnerable to the same kind of misbehavior. Second, both algorithms evaluate the trust of mesh clients based on their trust history as maintained by the mesh routers. The only major difference between the two algorithms is the trust evaluation process. MDA uses correlation, whereas CORIDS is a reward-based intrusion detection algorithm. All these features make it interesting to observe the behavior of CORIDS compared to MDA.

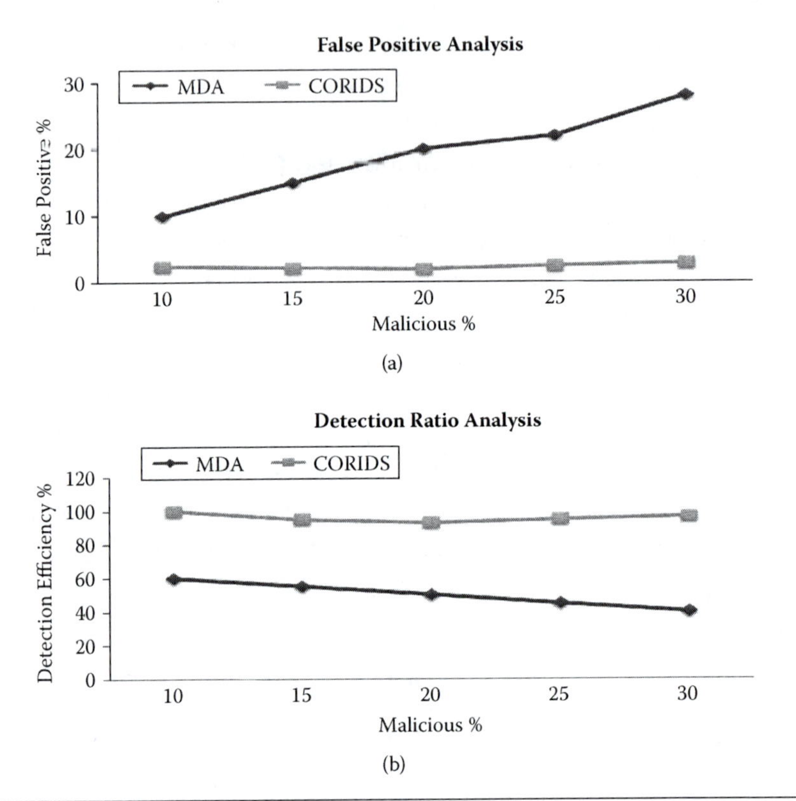

Figure 7.8 (a) Comparison of false positive analyses between CORIDS and MDA. (b) Comparison of detection ratio analyses between CORIDS and MDA. (From N. Deb, M. Chakraborty, and N. Chaki, 2013. *Security and Communication Network*, John Wiley & Sons. With permission.)

Figures 7.8a and b show the comparative results. The gray line shows the performance of CORIDS compared to that of MDA, as shown by the black line. The false positive percentage is extremely low (about 2%) and consistent for CORIDS. For MDA, the number of false positives is a linearly increasing function. Also, the false positive percentage is quite high, starting from 10% (Figure 7.8a). In terms of detection efficiency, CORIDS performs consistently better, with an efficiency of 90% and above. For MDA, the detection efficiency is a linearly decreasing function of the density of malicious nodes. Also, the maximum detection efficiency achieved by MDA is 60% (Figure 7.8b).

Thus, it can be concluded that CORIDS is a lightweight algorithm that uses very few control messages and executes from the backbone routers only. Control messages are restricted to each cluster. The percentage of false positives is relatively low for CORIDS and remains consistent with variation in node density and node mobility. There are no false negatives when CORIDS is used to detect DoS attacks. This is demonstrated by the fact that the detection efficiency of CORIDS remains consistently high for blackhole attacks and is 100% for DoS attacks.

7.5 Conclusions

The field of intrusion detection is increasing in importance. Service providers are trying to provide better quality services at affordable rates. Customers are increasingly exchanging sensitive data, including their credit card information or bank details, through online services.

Secure mobile applications are the need of the hour, and in this scenario CORIDS holds huge potential. The algorithm performs consistently well for increasing mobility, varying node density, and different percentages of malicious nodes. Thus, CORIDS holds promise to serve as the groundwork for advanced research in various domains.

Vehicular ad-hoc networks (VANETs) are one such application domain. One of the major obstacles faced by researchers while providing solutions for VANETs is the mobility of vehicles. In the simulation of CORIDS, the mobility was varied from 10 to 90 mps. The consistency of results reported above indicates that it can be effective in providing security in VANETs in an energy-efficient manner. Besides, it is often observed that customers cannot access the network when node population increases drastically during concerts or at stadiums. The scalability of CORIDS may solve this issue. In fact, the consistency in the performance of CORIDS makes it a candidate for any scalable application. In this chapter, a blackhole attack and a particular type of DoS attack were simulated, where a malicious node generated spurious packets and flooded them throughout the network. In the future, the current research may be extended to find the performance of CORIDS in different application domains and on actual test beds with a large number of nodes. Similarly, the performance of CORIDS may be tested for other types of DoS attacks, such as simple jamming of the channel by transmitting some random signal, etc.

References

1. Ian F. Akyildiz, X. Wang, and W. Wang, Wireless Mesh Networks: A Survey, *Journal of Computer Networks*, 47, 445–487, 2004.
2. Roberto Riggio, Francesco De Pellegrini, Daniele Miorandi, and Imrich Chlamtac, A Knowledge Plane for Wireless Mesh Networks, *Ad hoc and Sensor Wireless Networks*, 5(3–4), 293–311, 2008.
3. Roberto Riggio, Tinku Rasheed, Stefano Testi, Fabrizio Granelli, and Imrich Chlamtac, Interference and Traffic Aware Channel Assignment in WiFi-Based Wireless Mesh Networks, *Ad hoc Networks*, 9(5), 864–875, 2011.
4. N. Asoka and P. Ginzboorg, Key Arrangement in Ad hoc Networks, at Proceedings of the Fourth Nordic Workshop on Secure IT Systems (Nordsec'99), 1999.
5. Dan Zhou, *Security Issues in Ad hoc Network: The Handbook of Ad hoc Wireless Networks*, CRC Press, Boca Raton, FL, 2003.
6. S. Axelsson, *Intrusion Detection Systems: A Survey and Taxonomy*, Technical Report 99-15, Department of Computer Engineering, Chalmers University of Technology, Sweden, March 2000.
7. J. McHugh, Intrusion and Intrusion Detection, *International Journal of Information Security*, 1, 14–35, 2001.
8. R. Chaki and N. Chaki, IDSX: A Cluster Based Collaborative Intrusion Detection Algorithm for Mobile Ad-Hoc Network, in *International Conference on Computer Information System and Industrial Management Applications (CISIM'07)*, Minneapolis, MN, June 28–30, 2007, pp. 179–184.
9. Aikaterini Mitrokotsa, Nikos Komninos, and Christos Douligeris, Intrusion Detection with Neural Networks and Watermarking Techniques for MANET, in *Proceedings of IEEE International Conference on Pervasive Services*, 2007, pp. 118–127.
10. Noman Mohammed, Hadi Otrok, Lingyu Wang, Mourad Debbabi, and Prabir Bhattacharya, Mechanism Design-Based Secure Leader Election Model for Intrusion Detection in MANET, *IEEE Transactions on Dependable and Secure Computing*, 8(1), 89–103, 2011.
11. H. Yang, J. Shu, X. Meng, and S. Lu, SCAN: Self-Organized Network-Layer Security in Mobile Ad hoc Networks, *IEEE Journal on Selected Areas in Communications*, 24, 261–273, 2006.
12. Debdutta Barman Roy, Rituparna Chaki, and Nabendu Chaki, BHIDS: A New Cluster Based Algorithm for Black Hole IDS, *Journal of Security and Communication Networks*, 3(2–3), 278–288, 2010.
13. Abdul Hamid, Shariful Islam, and Choong Seon Hong, Misbehavior Detection in Wireless Mesh Networks, in *International Conference on Advanced Communication Technology*, 2008, pp. 1167–1169.
14. Z. Zhang, Farid Nait-Abdesselam, Pin-Han Ho, and Xiaodong Lin, RADAR: A Reputation-Based Scheme for Detecting Anomalous Nodes in Wireless Mesh Networks, in *Proceedings of the Wireless Communications and Networking Conference (WCNC)*, 2008, pp. 2621–2626.

15. Hua Ye and Mehran Ektesabi, RFIDS: Radio Frequency Indoor Intrusion Detection System, in *Proceedings of the World Congress on Engineering 2008, WCE 2008*, London, July 2–4, 2008, vol. I, pp. 401–404.
16. A. Lakhina, M. Crovella, and C. Diot, Diagnosing Network-Wide Traffic Anomalies, in *International Conference on Special Interest Group of Data Communication*, August 2004, pp. 219–230.
17. Fabian Hugelshofer, Paul Smith, David Hutchison, and Nicholas J.P. Race, OpenLIDS: A Lightweight Intrusion Detection System for Wireless Mesh Networks, in *International Conference on Mobile Computing and Networking*, Beijing, China, September 2009, pp. 309–320.
18. Devu Manikantan Shila and Tricha Anjali, Defending Selective Forwarding Attacks in WMNs, in *Proceedings of IEEE International Conference on Electro Information Technology, EIT 2008*, 2008, pp. 96–101.
19. Jaydip Sen and Kaustav Goswami, An Algorithm for Detection of Selfish Nodes in Wireless Mesh Networks, in *Proceedings of the International Symposium on Intelligent Information Systems and Applications (IISA'09)*, October 2009, pp. 571–576.
20. Meijuan Gao, Jingwen Tian, Kai Li, and Hao Wu, Community Intrusion Detection and Pre-Warning System Based on Wireless Mesh Networks, in *Proceedings of IEEE Conference on Robotics, Automation and Mechatronics*, September 2008, pp. 1066–1070.
21. D. Makaroff, P. Smith, N.J.P. Race, and D. Hutchison, Intrusion Detection Systems for Wireless Mesh Networks, in *Proceedings of the 5th IEEE International Conference on Mobile Ad hoc and Sensor Systems*, September 2008, pp. 610–616.
22. Z. Bankovic, D. Fraga, J. Manuel Moya, J. Carlos Vallejo, P. Malaga, A. Araujo, J. de Goyeneche, E. Romero, J. Blesa, and D. Villanueva, Improving Security in WMNs with Reputation Systems and Self-Organizing Maps, *Journal of Network and Computer Applications*, 34(2), 455–463, 2011.

8

Future Trends in WAN Security

TAPALINA BHATTASALI, MANALI CHAKRABORTY, AND NABENDU CHAKI

Contents

8.1 Overview of Future Trends in Wireless Ad-Hoc Network Security

This chapter focuses on the future trends in wireless ad-hoc network (WAN) security. It is sometimes very useful to predict the future to get new ideas and visualize the present in a more appropriate context. Future trends are the consequence of today's activities. There are many open issues related to the future of wireless ad-hoc networks. A scenario of completely unrestricted "anytime, anywhere" communications using this technology seems to be inevitable in the nearest future. Wireless networking can be very well suited for the next-generation communications. Wireless ad-hoc networks have the potential to change how the communication world is seen. One of the major concerns today is the aggressive marketing policies of corporations. The industrial houses are perpetually in the rat race of claiming edges over each other. This remains their prime interest, and the big companies are just fascinated to see themselves the winner at the cost of almost anything. Quite often this thrust of staging eye-catching marketing stunts and promising more flexible and powerful user applications prompts them to market technologies in premature states. Security-related problems are unavoidable in such a scenario where technology is not thoroughly tested and critically analyzed.

Section 8.1 gives a brief overview of future trends in WAN security. Section 8.2 presents the idea of secure cloud services in the wireless ad-hoc network environment. It briefly describes a security architecture of cloud services on WAN to solve the problems associated with an integrated cloud-WAN environment. Section 8.3 includes ideas about secure smart grid architecture in a limited-energy wireless ad-hoc environment. Section 8.4 presents an idea about energy-efficient intrusion detection in WAN having a high probability of being compromised. Finally, Section 8.5 concludes the chapter.

In the next generation of the network, technology needs to be adapted to support the increase in network traffic being driven by an alarming number of devices and enhanced demand of huge bandwidth for big data. It has been seen that the total number of devices grew from 500 million to 1.2 billion in 2012 [26]. There is still space for improvements since the performance of WAN is typically poor compared to that of other wireless technologies. In the field of wireless ad-hoc networks, new application areas are emerging, such as cloud services, smart grid applications, energy-efficient applications, vehicular ad-hoc networks (VANETs), etc. VANET is based on ad-hoc connections between vehicles to improve safety in transport. With the advent of novel applications, it is expected that VANET will be widely adopted to provide a set of services that can also be used for critical applications efficiently. Sensor network area is also related to ad-hoc networks that move toward the future generation of Internet of Things applications. The wireless mesh network (WMN) is also a related area in this domain.

This chapter focuses mainly on securing cloud services and smart grid architecture on WAN. Another considerable issue is energy-efficient computation in this type of network. The real challenge lies in finding the solutions to handle high volumes of data securely while maintaining quality-of-service requirements. An optimized solution of future WANs can control the Internet to reshape the way we use critical applications. To achieve greater efficiency, we require solutions for network traffic from limited-resource devices, cloud services, new types of IPv6 content, etc. An additional burden is placed on WAN in terms of increased demand for security.

In order to manage network security effectively, one should gain full visibility into how network capacity is being used by the applications. Besides understanding what type of traffic is flowing through the network, bandwidth usage per application should also be measured. The most critical question is how to improve performance without investing in new network infrastructure or upgrading bandwidth. Cloud service has a profound impact in this regard. As more applications are deployed through cloud services, it is more efficient from the cost and latency perspective. This concept reduces network costs and improves performance. The privacy and security challenges are also increasing day by day, which include concern about trust establishment, secure transmission, protecting data privacy, ensuring data integrity, identifying the most dangerous attacks, and designing solutions of intrusion detection.

Smart grid technology also places greater demands for reliability on WAN communications [3]. This will allow prioritized communication—high priority for abnormal events and system control operations and low priority for asset management tasks. Connectivity protection and its data confidentiality in smart grid applications in the WAN environment are critical. In this scenario, having a fast and real-time reaction upon an abnormal event is vital. The wireless architecture should aim for high-priority, low-latency alerts when abnormality occurs. In a smart grid almost all the nodes are fixed, so the communication architecture does not consider node mobility explicitly. In this environment, high overhead is created by multiple nodes trying to send the same information and excessive use of control packets. This reduces available bandwidth for data traffic, and can also result in higher latencies for critical alert packets. Therefore, it is necessary to ensure that overhead be kept low.

8.2 Securing Cloud Services in Wireless Ad-Hoc Network

In the last section, a brief overview of the future trends in WAN security was given. A detailed account on securing cloud services on WAN is presented in this section. The boundary between the physical world and the digital world has been dissolved due to advances in the areas of ubiquitous computing. WAN collects data about the physical environment. However, collected data cannot be processed over long periods of time due to lack of storage capacity. In WAN, all the nodes and topology are unstable due to the fact that there is no decentralized location where all the shared information can be stored with knowledge of the network resource. This will also lead to a great challenge in the quality of service (QoS) of this type of network. Cloud computing provides an alternative for data storage and computation. Ubiquitous ad-hoc environments and cloud computing complement each other. Cloud service is used to provide resources in on-demand environments. It makes response time faster and cost lower. The U.S. National Institute of Standards and Technology (NIST) [25] defined cloud computing this way: "Cloud computing is a model for enabling convenient, on-demand network access to a shared pool of configurable computing resources (e.g. networks, servers, storage, applications, and services) that can be rapidly provisioned and released with minimal management effort or service provider interaction."

Because of the distributed nature, achieving security for cloud environments is consistently raised as a major concern. In the current scenario, where cyber attacks and data leakage incidents are increasing, it must be ensured that data assets are well protected when they

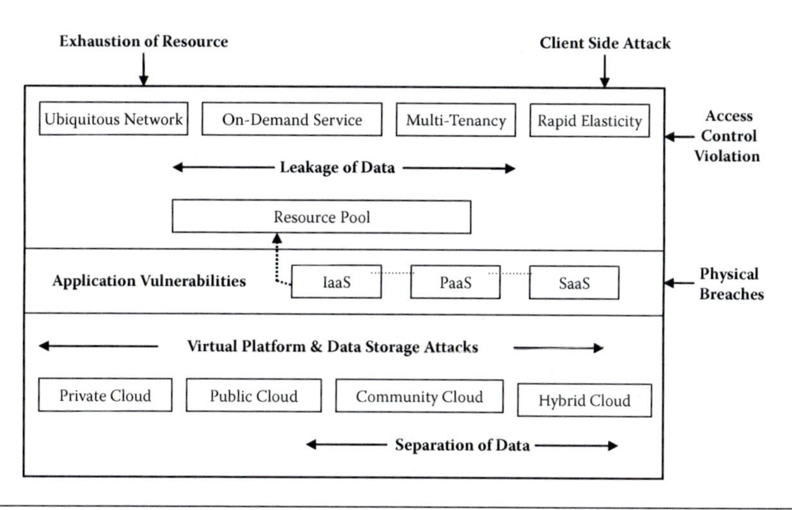

Figure 8.1 Threat model in cloud environment.

are kept in the hands of a third party. The data stored in the cloud are accessed a large number of times and are often subject to different types of changes. This may comprise bank accounts, passwords, and highly confidential files not to be read by someone other than the owner. Hence, even a small slip may result in loss of data security. Cloud security mainly deals with identity management, encryption, intrusion detection, forensics issues, and risk assessment, along with the responsibility for deciding how and where data are stored and accessed in the cloud. It has been identified that most of the potential risks include malware, data leakage or breach, denial-of-service (DoS) attacks, etc. Figure 8.1 represents the basic threat model in a cloud environment. It mainly categorizes the attacks as denial-of-service attacks, client-side attacks, data leakage, violation of access, application vulnerabilities, physical breaches, data separation, virtual platform attacks, and data storage attacks on the cloud environment that includes infrastructure-as-a-service (IaaS), platform-as-a-service (PanS), software-as-a-service (SaaS).

In order to secure the cloud, security options must be analyzed to make sure data protection is in the right place. Therefore, focus must be on how to balance cloud computing security risks with the convenience in WAN. Before uploading data to a cloud from any of the wireless hosts, security policies need to be included. The information in the cloud database will be used to detect vulnerabilities in the data sent from wireless nodes. This will be done in real time, to ensure that information is delivered in a secure manner, when it is needed. A secure cloud service architecture on WAN is presented next to minimize the problems that arise in this scenario. Security issues and attacks in a cloud are different for different layers of the underlying networking infrastructure, such as network layer, application layer, or from the host level. Some of these attacks are listed below.

8.2.1 *Attacks on Cloud Computing Systems*

Chapter 4 presented a thorough discussion on the different types of attacks for computer networks. As any cloud architecture is built on underlying network connectivity, the common networking threats, like the man-in-the-middle attack or denial of services, are highly relevant for the cloud as well. In this section, we present a few common attacks, typical for the cloud domain:

SQL injection attacks: In this type of attack, a hacker can access a cloud database inserting a malicious code in the Structured Query Language (SQL) code for a standard query [17]. This not only allows the attacker to access sensitive data, but also may create confusion by inserting wrong data in the cloud database.

Cross-site scripting (XSS) attacks: With this attack, the intruder injects malicious scripts in web contents [18]. Static websites don't suffer from the XSS attacks. Cross-site attacks are planned for the dynamic websites providing diverse and on-the-fly services to the users. There are two variants of this attack: stored XSS and reflected XSS. In a stored XSS, the attacker stores the malicious code in a resource managed by the web application. The actual attack is triggered at a stage when the victim requests a dynamic page that is constructed from the contents of this resource. However, in case of a reflected XSS, the attack script is not stored in the web application. Such an attack is immediately reflected back to the user.

Reused IP address attack: This is a typical network attack. We know that each node of a network is provided an IP address that has a specific range, depending on the type of network. This attack is somewhat analogous to a situation with two successive bank ATM users, when the second user finds that although the previous user has left, his session is still in use. From the user's perspective a major difference is that in the case of an ATM user, the security of the first ATM user is at stake. In the reused IP address attack in a network, the privacy of the second user may be compromised.

When a particular user A moves out of a network, then the IP address so far associated with A is assigned to a new user, say B. This sometimes risks the security of the new user, as there is a certain time lag between the change of an IP address in the Domain Name System (DNS) and the clearing of that address in DNS caches [19]. Hence, it may be said that sometimes, though the old IP address is being assigned to a new user, the chances of accessing the data by the old user exists. This is because the address still exists in the DNS cache and the data belonging to B may become accessible to A, violating the privacy of B.

Sniffer attacks: Sniffing refers to unauthorized reading of data packets flowing in a network that are not encrypted. Thus, an attacker can capture vital information flowing across the network. A sniffer program works in the promiscuous mode to track all data flowing in the network [20].

Google hacking: Google hacking refers to using the Google search engine to find sensitive information that a hacker can use to his benefit toward hacking a user's account. The hackers find the security loopholes of a cloud infrastructure or a cloud-based system using Google. The attacker may even use Google to find the target with the loopholes and containing the right kind of data or service that is being targeted. After gathering the necessary information, the system is hacked. In a well-known Google hacking event of the recent past, the login details of various gmail users were stolen by a group of hackers [21].

Account hijacking: This is one of the most serious threats for any commercial cloud service provider. According to cloud security alliance (CSA) [22], account and service hijacking often occur using credentials stolen from genuine users. With stolen credentials, attackers can often access critical areas of deployed cloud computing services, allowing them to compromise the confidentiality, integrity, and availability of those services.

Abuse of cloud services: More and more paid and unpaid services are being deployed using cloud. Very important government to citizen (G2C) services are increasingly being deployed as cloud services. Hackers are taking advantage of these and often use the immanence power of cloud computing to hack other services. This is abuse of existing cloud services. The impact of such an attack goes far beyond the Google hacking discussed above.

CAPTCHA breaking: Internet users are often asked to enter some text displayed in a box where characters are oriented in all possible angels. These are CAPTCHAs. Free mail services like Google, Yahoo, and a large number of other websites use CAPTCHA to prevent usage of internet resources by robots or computers. Even the multiple website registrations, dictionary attacks, etc., by an automated program are prevented by using a CAPTCHA.

However, recently it has been found that spammers are able to break the CAPTCHA [23] provided by popular free mail service providers. Various techniques, such as implementing letter overlap, using variable fonts, increasing the string length, and using a background, are being tried to secure CAPTCHAs [24].

Because it is known to all that integrated WAN-cloud communication imposes hard real-time requirements, this architecture must not introduce long delays. To investigate service delay, cloud should collect node and WAN status information and predict appropriate actions to be taken; cloud should serve as a server, i.e., assisting a mobile node to establish trust with another node controlled in different domains; and cloud should emulate the actions of the network for post-event analysis. From the above discussion, it has been seen that cloud service has great potential to bring more application scenarios securely on wireless ad-hoc networks.

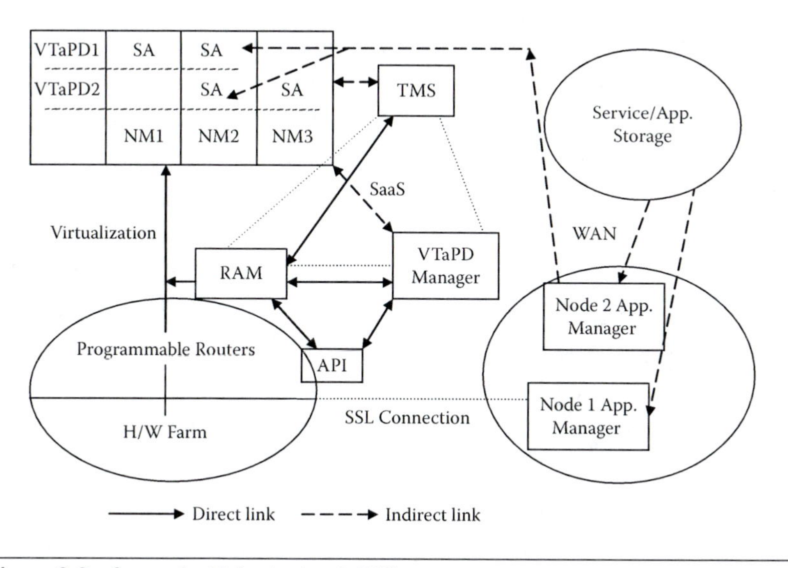

Figure 8.2 Secure cloud infrastructure in WAN.

8.2.2 An Architecture for Secured Cloud Service

Figure 8.2 shows the conceptual infrastructure to secure cloud services in WAN [2]. This is basically designed to secure data access policy management for protecting users' data, to monitor WAN status for risk assessments, to detect intrusion and respond accordingly, and to simulate and predict future WAN status for decision making. This also provides trust management and feedback capability to the users. Trust management includes identity management, key management, efficient data access control, risk assessment, etc., to provide security as a service (SeaaS), which can offer security service according to the request from different applications.

This model mainly depends on several components, such as virtual trusted and provisioning domain (VTaPD), software agents (SAs), programmable router, node manager (NM), resource and application manager (RAM), and trust manager server (TMS). VTaPD service is used to isolate information flows from different security domains through programmable routers. Software agents are used to link the cloud services and wireless devices. Each device can have multiple SAs for different services, which are managed by the application manager of the device. The application interface provides interfaces to the VTaPD manager and RAM, which constructs VTaPDs according to the direction of the VTaPD manager and TMS. The VTaPD manager collects context awareness information and uses it for intrusion detection and risk management. TMS acts as trust authority, which handles attribute-based key distribution and revocation. It provides an identity search for devices belonging to multiple domains and policy checking to provide a unified trust management system. This framework considers time synchronization service on wireless devices and virtual routing domain to emulate the routing behaviors of the WAN and communicate the decisions to the nodes. This integrated framework of secure cloud services in WAN will reduce the uncertainty by functioning as information storage. But there are several issues that need to be addressed in the near future. The first issue is whether this framework can protect users' data, even if the devices are compromised. To develop an efficient many-to-many secure group communication system, a fine-grained data access control mechanism μVTaPD needs to be constructed, where μ is used to specify different types of constraints. The next issue is how to construct and delete μVTaPD.

8.3 Smart Grid Security in Wireless Ad-Hoc Networks

The conventional electrical power grid that has been used for decades has met our needs in the past. However, as our society advances technologically, so do the expectations from various infrastructures surrounding us. Smart grid is an initiative to completely restructure the electrical power grid to meet the current and future requirements of its customers. Updating our electrical power grid could introduce new security vulnerabilities into the system. Therefore, security is one of the important aspects in smart grid technology.

A smart grid is an intelligent electricity network that integrates the actions of all users connected to it and makes use of advanced information, control, and communication technologies to save energy, reduce cost, and increase reliability and transparency. The easiest way to define the smart grid is by its characteristics. The smart grid is an upgrade to the current electrical power system, so it has all of the functionality of our current power system plus several new functionalities. These new functionalities cause more vulnerability to the system [33].

Smart grid is mainly composed of six basic systems: power generation system, distribution system, transmission network, data

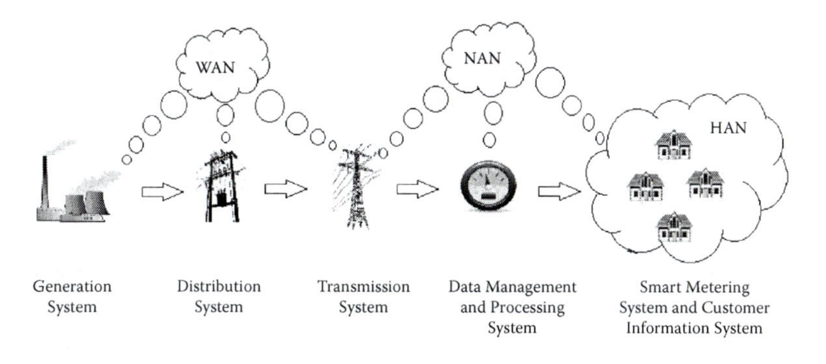

Figure 8.3 Smart grid architecture.

management and processing system, smart metering system, and customer information system. The network architecture of smart grid hierarchically consists of three components: home area network (HAN), neighborhood area network (NAN), and wide area network (WAN) [44], as shown in Figure 8.3. The HAN provides the communication between the smart meters in a home and other appliances in that home, while the NAN connects smart meters to the local data management and processing centers, and WAN provides access between the generation plants, distribution points, and transmission networks. The generation system, distribution points, and transmission networks build the core utility system of a smart grid.

Like the characteristics of each tier network, different wireless communication techniques can be adapted, i.e., WiFi or Zigbee for HAN in indoor small areas, WiMAX or WiFi for NAN with wireless mesh topology, and WiMAX, 4G, or cognitive radio for WAN [44, 45].

8.3.1 Security in Smart Grid

As smart grid technology is different from normal power grid technology, the security challenges in smart grid are also different from normal power grid technology. Beyond the application of traditional information technology (IT) security mechanisms, such as proper authentication, secure protocols, intrusion detection/response systems, and proper security engineering processes, security in the smart grid also faces novel challenges. Thus, the existing security solutions need to be upgraded, and also some new security solutions are needed for securing smart grid technology. This requires guaranteeing the stability of control systems that are also undergoing malicious disturbances [34, 35]. At the same time, IT security must take into account the real-time and analog nature of the grid and adapt risk management as graceful degradation (i.e., a slower, controlled, safe failure), as opposed to a sudden, disastrous failure when under attack.

A smart grid electric power system delivers electricity from producers to consumers using two-way smart meter technology that can remotely control consumer electricity use. This can help utilities conserve energy, reduce costs, increase reliability and transparency, and make processes more efficient. However, the increasing use of IT-based electric power systems increases cyber security vulnerabilities, and this increases the importance of cyber security. The main objective of providing security in smart grid is to maintain three important qualities in it: availability, integrity, and confidentiality.

Availability is the most important security objective. Smart grid is a critical real-time system and continuously monitors the state of the electrical power grid, and a disruption in communications can cause a loss of power. Thus, availability of the electrical power grid is its most important factor. By extension, the most important security object of most of the electrical power system components is also availability [33].

Integrity is the next important security objective in the smart grid. The smart grid uses data collected by various sensors and agents. These data are used to monitor the current state of the electrical power system. The integrity of these data is very important. Unauthorized modification of the data, or insertion of data from unknown sources, can cause failures or damage in the electrical power system.

The final security objective is confidentiality. There are certain areas in the smart grid where confidentiality is more important. Examples include the privacy of customer information, general corporation information, and electric market information.

Security of the smart grid can be divided into three categories: physical security, data security, and cyber security [39]. Physical security relates to protection of the smart grid's physical infrastructure, including advanced meter interface (AMI) hardware, such as smart meters, transmission lines, generating equipment, and control rooms, from damage. Such damage can be the result of intentional attacks using electromagnetic pulses or other weapons, or unintentional as the result of damage from electric storms. Data security refers to the privacy of the information that is transferred over the smart grid; it relates to customer information such as personal details, financial information, and energy usage patterns that can be misappropriated by hackers to do damage to individuals. Cyber security relates to the vulnerability of the grid to intentional infiltration by hackers using the Internet or other digital information management systems with the intention of disrupting the normal operation of the power delivery system.

The backbone of the smart grid will be its network. This network will connect the different components of the smart grid together, and allow two-way communication between them. Networking the components together will introduce security risks into the system. Two-way communication has the potential to create a new avenue for cyber attacks to reach the bulk power system and cause serious damage to this critical infrastructure by way of a customer's smart meters and other grid-connected smart technology. An attacker who gained access to the communication channels could order metering devices to disconnect customers, order previously shed loads to come back online prematurely, or order dispersed generation sources to turn off during periods when a load is approaching generation capacity, causing instability and outages on the bulk power system. Thus, to prevent these kinds of attacks, we need to secure the routing protocols in smart grid networks and also implement some robust intrusion detection system for a second line of defense [37].

Also, the smart meters are one of the weakest links in the smart grid security chain. Smart meters may be used by hackers as entry points into the broader power system. Hackers could hack into smart meters to take command and control of the advanced metering infrastructure, allowing a mass manipulation of services to homes and businesses [36, 41].

Cyber security must address not only deliberate attacks launched by disgruntled employees, agents of industrial espionage, and terrorists, but also inadvertent compromises of the information infrastructure due to user errors, equipment failures, and natural disasters. Vulnerabilities might allow an attacker to penetrate a network, gain access to control software, and alter load conditions to destabilize the grid in unpredictable ways [42].

However, already there exist a lot of secure routing protocols and intrusion detection systems for ad-hoc networks. Then why do we need some extra effort to secure the smart grid? The answer is because of the unique characteristics of smart grid technology, which differ from those of both traditional power grid systems and traditional ad-hoc networks.

We can also imagine the smart grid network as an ad-hoc network. Then it also implies that the existing security solutions for an ad-hoc network can be used for providing security in smart grid. However, there are some problems.

First, the nature of the network of smart grid is extremely large. For instance, it could be the case where 100,000 nodes (meters) generate meter traffic data every 10 min. And then this huge amount of data is analyzed to generate bills and to monitor the whole network. As a result, we have to incorporate scalability and reliability in the existing solutions so that this data can be delivered to the central utility control safely and in a timely manner [39].

Second, the traffic in a smart grid network will be traversing different types of networks, using a variety of media, ranging from fiber optics/broadband (e.g., for meters to base control center networking) to Zigbee/wireless local area network (WLAN) (e.g., for home networking). So interoperability is another key issue. It can be envisaged that in a complex system such as smart grid, heterogeneous communication technologies are required to meet the diverse needs of the system. Therefore, in contrast to conventional security solutions, the standardization of communications for smart grid means making interfaces, messages, and workflows interoperable. Thus, we can construct a totally new architecture for this kind of network, or we can combine different architectures for different layers, and build some interfaces to connect them with each other, so that they can communicate within themselves [36].

Besides, the traffic that will be generated by e-energy type applications in smart grid will likely be quite different from the traditional browsing/downloading/streaming applications that are in use today, with a mix of both real-time and non-real-time traffic being generated and distributed across different parts of a smart grid. Thus, the traditional security solutions may need to be revisited. The existing routing policies will also need to be changed to route real-time and non-real-time data simultaneously, with improved QoS [43].

8.3.2 *Some Possible Threats for Smart Grid Network*

There are some basic security threats [45] for the smart grid network:

- Bill manipulation/energy theft: An attack initiated by a consumer with the goal of manipulating billing information to obtain free energy.
- Unauthorized access from the customer endpoint: Compromising smart meters and other customer end devices to gain unauthorized access to the network.
- Interference with utility telecommunication: Unauthorized access to the core utility system, i.e., generation, distribution, and transmission system, causing mass power disruption.
- Mass load manipulation: Unauthorized access to distribution points and transmission networks, causing havoc in load manipulation.
- Denial of service: Jamming the network channels, requesting false demands, causing denial of service, etc.

8.3.3 *Research Challenges*

The smart grid is a large and complex system. Because of this complexity, research work typically only focuses on a single component. The different categories are listed below [33, 40, 46]:

- Security in smart meters
- Home area network security
- PCS security
- Security in distribution systems
- Cyber security in transmission network
- Smart grid communication protocol security

8.3.3.1 Smart Meter Security Smart metering is considered the first point where smart grid begins. The primary mission of a meter is to monitor power consumption. Smart meters are an electronic version of the power meters that are currently used. The electrical power readings are sent back to the power suppliers at regular intervals [41].

The security of smart meters is important because altered readings from the device can lead to incorrect billing and false power usage approximations. Altering smart meters can provide attackers with monetary gains, and since the device is installed at a customer's site, access to these devices is readily available.

8.3.3.2 Home Area Network Security The home area network (HAN) [46] is where the smart grid connects with the consumer. It is the part inside the home or place of business, and it is the part over which a utility or other service provider has the least control.

HAN security is clearly an important approach because the use cases and architectures are still new and evolving. Without understanding the architecture and the type of security vulnerabilities, it is difficult to build security systems. Additionally, the HAN security solution must take into account the scalability factor of the network and also the cost of implementation, because every home and business organization would potentially use the solution.

8.3.3.3 PCS Security Process control systems (PCSs) are the components responsible for monitoring and controlling physical properties of the electrical power grid. The PCSs in smart grid will be monitoring large geographical areas of the power grid. This means that there will be many entry points to get into the network. PCSs used in the smart grid will need to address these security issues.

There already exist some works addressing different issues, like smart meter intrusion detection systems (IDSs), redundant readings, and privacy. But they are not sufficient. The IDSs are generally signature based, so they cannot detect new attacks. One method to verify the accuracy of smart meters is to install a separate electrical energy measuring device that compares its reads to the readings that the power supplier received from the smart meter. The problem with this approach is that it introduces confidentiality risks. Attackers can intercept the data used to verify the integrity of the smart meter. So we need to use some kind of encryption system to secure smart meters.

8.3.3.4 Security in Distribution Systems The primary goal of a distribution system is to improve power delivery system reliability, performance, and quality. The present distribution networks have many visible single points of failures, making service disruption due to cyber or physical attack a serious risk [46].

With smart grid distribution systems, outages are identified and located in real time. This allows rapid deployment of the resources to the right location to resolve problems. Distributed generation, automated switching, and self-healing capabilities are used for better functionalities. But this also makes the distribution system vulnerable to many security threats.

8.3.3.5 Cyber Security in Transmission Network The power grid connects power through a series of substations. The totality of this represents the transmission network that is used to transmit power. To secure a transmission network, the following qualities have to be maintained [46]:

Self-healing: This will ensure that when transmission is affected, the system will automatically take corrective measures.

Power quality: If the power quality is high, then the transmission system can provide high-quality service. However, if it is low, then there might be a problem.

Energy storage: If generation is taken out, then other stored energy will be available.

8.3.3.6 Smart Grid Communication Protocol Security The smart grid communication protocols are the next category of smart grid security research. The smart grid relies on communication between its different components in order to function. Each of the components has different communication requirements. The communication requirements range from very low latency to high data throughput, and each has a set of security needs [40].

The smart grid will need several communication protocols to meet the varying connection requirements. The security of smart grid communication protocols is important because the network communication is the backbone of the smart grid. Many of the major smart grid functionalities cannot take place without communication. The security objectives that are important depend on which components are communicating, and what data they are exchanging.

Smart grid communication protocol security is a challenge because there are many different components communicating, each with their own set of communication requirements. Another issue is that the smart grid technology needs to integrate with legacy power systems, and many of these devices have constraints that must be considered. Legacy devices can typically introduce security vulnerabilities into the system because of a lack of security support.

8.3.4 Conclusion

Therefore, we can see that smart grid is a new frontier for communications and networking research. It poses many unique challenges and opportunities, e.g., interoperability, scalability, and security. The success of future smart grid depends heavily on the communication infrastructure, devices, security and enabling services, and software. Although there has been a lot of work toward the security in smart grid, some issues still need to be addressed. It is required that we build a secure architecture with a secured data analysis system that can sustain a certain level of physical and cyber attacks, besides maintaining the basic characteristics of smart grid, i.e., availability, integrity, and confidentiality.

8.4 Energy-Efficient Intrusion Detection in WAN

The previous section provides a basic idea about smart grid security in the context of WAN. A brief review of energy-efficient intrusion detection [8] in WAN has been presented here. Wireless networks are more vulnerable to attacks than wired networks. In this type of network, malicious nodes will be able to join the network at any time because of its infrastructure-less nature. Ad-hoc wireless networks with their changing topology and distributed nature are more prone to intrusions [15]. Therefore, a need to quickly detect and isolate malicious nodes or networks arises. Securely distributing information about malicious entities in the presence of an intruder is a big challenge. Avoiding malicious entities on top of maintaining connectivity is another challenge. As the demand for wireless networks grows day by day, intrusion detection becomes of high importance [29]. Wireless ad-hoc networks are more vulnerable to intrusions from any direction. Each node in the network must be aware to deal with the intruders. It is difficult to track a single compromised node in a large network, because attacks from compromised nodes are much harder to detect. Ad-hoc networks may rely on cooperative participation [27] of the members within a decentralized architecture. Intruders can take advantage of this lack of centralized architecture to launch new types of attacks. It is known that building such ad-hoc networks poses significant technical challenges because of the many constraints imposed by the environment. As nodes are generally battery operated, they need to be energy conserving [9]. Therefore, any operation in this field must be lightweight to maximize battery life [30]. Several technologies are being developed to achieve the goal of optimized energy consumption, even in the case of intrusion detection.

Intrusion is defined as any set of actions that generally attempt to compromise availability, integrity, and confidentiality of a network resource. Since prevention techniques may not be sufficient and new intrusions continually emerge, IDS is a necessary component of a security system. An IDS is used to detect possible violations of a security policy by monitoring system activities. In order to identify either an outside intrusion or an inside intrusion [10], IDSs normally perform the following tasks: monitoring the network, analyzing collected data, identifying intruders, generating alarms, and tracking intruders to prevent such attacks in the future. These functionalities are encapsulated in several components, like data collector, data storage, data processor, and detection engine, all of which are controlled by the system configuration components.

It is known that intrusion detection methods are classified into three main techniques: anomaly based, misuse based, and specification based. An anomaly-based technique creates a profile of normal behaviors. It detects anomalies when recorded behavior deviates from normal behaviors. Misuse-based detection compares known attack signatures with current system activities. It is efficient and has a low false positive rate only for known attacks. Both anomaly-based and misuse-based approaches have their strengths and weaknesses. The specification-based technique is introduced as an alternative that combines the strengths of anomaly-based and misuse-based detection techniques, providing detection of known and unknown attacks with lower false positive rates.

Due to the decentralized nature of a wireless network, the main focus is on distributed solutions of intrusion detection for the network. Energy-aware design and evaluation of the intrusion detection system [28] for WAN require in-depth practical knowledge of energy consumption behavior of actual wireless devices. But very little practical information is available about the energy consumption behavior. Wireless devices normally operate for a long period of time, depending on their battery energy. Therefore, energy awareness is a major concern in wireless networking. To minimize energy consumption, one consideration should be to minimize the total energy needed for intrusion detection [5], and another consideration should be to look at the methods that extend the battery lifetime of the nodes. The energy consumption of the network interface can be significant, especially for smaller devices. It is sometimes assumed that bandwidth utilization and energy consumption are almost synonymous. In some cases, energy is often treated for purposes of minimizing cost or maximizing time to the network partition. Therefore, to design an energy-efficient intrusion detection system in WAN, issues like accuracy, energy consumption, and real-time response need to be considered.

Several intrusion detection systems have been proposed to deal with the problem of intrusion in wireless networks, some of which are extended versions of IDSs in wired networks. Energy awareness in wireless ad-hoc networks becomes a major issue when considering intrusion detection in larger networks. Monitoring intrusive activity normally occurs from either host-based IDSs or network-based IDSs. Beside this, hybrid intrusion detection systems incorporate multiple features into a single system. These are generally based on agents [11–13] who move throughout the network to provide an effective solution. Energy efficiency is one of the most important considerations in wireless devices due to the limitation of the battery life. Here an energy-efficient hybrid intrusion detection system (EEHIDS) is briefly discussed and compared with existing system power-aware agent-based intrusion detection (SPAID) [4] for performance evaluation.

8.4.1 Energy-Efficient Hybrid Intrusion Detection System (EEHIDS)

A hybrid agent-based intrusion detection system, EEHIDS is used to detect intrusion in an energy-efficient way [14], [16]. It is used to determine the duration for which a particular node can monitor network status. It focuses on the available energy level in each of the nodes to determine the nodes that can monitor the network. Energy awareness in the network results in maintaining energy for network monitoring by determining energy drainage of any node. The advantage of this approach is its inherent flexibility. Only fewer nodes are eligible for becoming candidates of network monitors. EEHIDS is built on an agent-based framework. It includes the following agents to perform its functions.

Network monitor: Only a limited number of nodes will have sensor agents for the network packet monitor. The main focus is to preserve the total computational energy and battery energy of hosts.

Host monitor: Every node on the network will be monitored internally by a host monitor agent. This includes both system level and application level monitoring.

Decision maker: Every node will decide the intrusion threat level. Certain nodes will collect intrusion-related data and make final decisions.

Actor: Every node will have an actor that is responsible for solving the intrusion status of a host.

There are three types of major agents categorized as monitor, decision maker, and actor agents. Some of them are present on all hosts, while others are distributed to only a selected group of nodes. In WAN, the elected network monitor nodes will include decision maker and actor modules. Functionalities must be distributed efficiently to save resources. Decision maker agents consider the energy metric, namely, network monitoring energy estimation (NeMEE). It is a node-specific metric to estimate energy consumption per node for running the network monitor agent. The NeMEE metric considers the average number of wireless links, used wireless protocol, remaining battery energy, etc. It can also estimate the duration the node remains at the same energy level without refreshment. The calculation of the parameter NeMEE involves calculating the duration for which the node can continue as a network monitor, along with its normal operations. NeMEE is calculated as shown in Equation (8.1):

$$\mathbf{NeMEE' = TBER/TEC_{mon}} \tag{8.1}$$

where TBER is the total battery energy remaining at the instant of node selection and TEC_{mon} is the total energy consumption with the network monitor node.

$$\mathbf{NeMEE' = TBER/TEC} \tag{8.2}$$

In the absence of measurement for energy consumption of the network monitor, NeMEE is assumed as NeMEE′. The value of NeMEE′ is directly available from most distributed wireless networks. TEC is the total energy consumption before the node is selected for network monitoring. Like SPAID, EEHIDS also considers a multihop network for selection of the network monitor within a cluster. The advantage of this type of node selection is that it allows complete coverage of all nodes and links in a network, but it creates redundancy in intrusion detection data collection. EEHIDS is an energy-efficient variation of SPAID. The EEHIDS approach considers each of the initially allocated monitors and the nodes they monitor to be a single tree. The network monitor node is treated as a root, and the nodes being monitored as its child. The root node and its child nodes form individual clusters. As a result, network topology gets divided into clusters in a tree-like fashion, only for intrusion detection purposes. After such cluster formation, when any drainage in energy levels takes place to the monitors, any other child node having higher battery energy is selected as a network monitor of that cluster. Only a limited number of clusters is kept active for a certain period of time. It is feasible that the monitor node gets rearranged within the cluster. The node selection process could be considered for the whole network only when no single node within a cluster has enough potential to monitor the network, or when a new node with a higher NeMEE value enters the existing network. In Figure 8.4, the EEHIDS algorithm is presented in brief.

EEHIDS Algorithm

Step 1: Set a constraint on the NeMEE value of nodes which are allowed to compete for becoming a network monitor node.

Step 2: Organize different nodes in increasing values of NeMEE, for all nodes that satisfy the NeMEE constraint.

Step 3: Initially set hop radius to 1 and increment for each insufficient node selection with the current hop radius.

Step 4: Consider node selection incrementally, starting from the first node having highest NeMEE value to the set of all nodes in the network by incrementing one node each time. This set is known as the working set (WoS) of nodes.

Step 5: Voting for network monitor node selection, considering the limitation that only WoS participants are eligible for being candidates.

Step 6: Check acceptability of nodes. If all are not represented by the set of selected nodes, then WoS is expanded and it is repeated from Step 4. If WoS equals the NeMEE ordered list, then increment the hop radius, and is repeated from Step 3.

Step 7: Create individual tree-structured clusters by considering nodes selected as network monitors as roots and nodes being monitored as child nodes.

Step 8: Changes in energy levels of the root nodes in each cluster will be informed to the child and voting takes place within the cluster to form a new monitor node.

Figure 8.4 Key steps of EEHIDS algorithm.

Steps 1 to 6 are similar to those in SPAID. The difference is that the steps of EEHIDS are also suitable for highly dynamic networks.

8.4.2 Intrusion Detection in WAN

Here, detection of intrusions in the network is done with the help of cellular automata (CA). It could classify a packet transmitted through the network as either normal or compromised. The use of CA is helpful in the identification of well-known intrusions as well as new intrusions.

Intrusion detection using an agent framework depends on both local response and global response. However, individual cells in CA can only communicate locally without the existence of a central control. The main idea behind using a cellular automata framework is to understand how it is developed using genetic algorithms to perform computational tasks requiring global information processing. Therefore, this framework provides an appropriate approach for intrusion detection where dynamic systems are created to provide local information processing as well as coordinated global information processing.

To solve problems, localized structures are used by cellular automata. Genetic algorithms are normally used to identify populations of candidate hypotheses to a single global optimum [7]. For this reason, a set of rules needs to be considered for designing efficient IDSs. It is not possible to detect whether a network connection is normal or anomalous accurately by using only one rule. Multiple rules are required to detect unrelated anomalies. CA represent a generalized linear classifier looking for the maximum margin hyperplane between two classes in the feature space. The optimal position of the class boundary is obtained as a linear combination of some training samples that are placed near the boundary itself. The hyperplanes are defined by the following set of linear equations [7]:

$$\mathbf{w.x + b = 0, w \in R^d, b \in R} \tag{8.3}$$

Each input x is subject to the decision function O(x), where

$$\mathbf{O(x) = sign\ (w.x + b)} \tag{8.4}$$

$$\mathbf{Output,\ y = +1,\ if\ w.x + b \geq 1} \tag{8.5}$$

$$\mathbf{Output,\ y = -1,\ if\ w.x + b \leq -1} \tag{8.6}$$

The margin width of the hyperplane is calculated by considering the plus plane and minus plane. The plus plane is represented by Equation (8.7):

$$\mathbf{x: w.x + b = +1} \tag{8.7}$$

The minus plane is represented by Equation (8.8):

$$\mathbf{x: w.x + b = -1} \tag{8.8}$$

The perpendicular distance of the plus plane from the classifier boundary is in Equation (8.9):

$$|\mathbf{1-b}|/||\mathbf{w}|| \tag{8.9}$$

The perpendicular distance of the minus plane from the classifier boundary is in Equation (8.10):

$$|\mathbf{-1-b}|/||\mathbf{w}|| \tag{8.10}$$

The margin width can be calculated by Equation (8.11):

$$|\mathbf{1-b+1+b}|/||\mathbf{w}|| = \mathbf{2}/||\mathbf{w}|| \tag{8.11}$$

In order to determine the maximum margin between the pair of hyperplanes, the value of w needs to be minimized. The EEHIDS flow is as follows:

Query formation → Intrusion attributes → Presented features → Comparison
Training data → Intrusion attributes → Presented features → Comparison

If new intrusion is detected after comparison, it is added to the training set; otherwise, necessary steps are taken to prevent intrusion

The two phases in EEHIDS to detect intrusion using cellular automata [31, 32] are:

1. Training phase: In this phase, an intrusion database is populated with sample intrusions and their attributes. Optimal hyperplanes for each of the binary classifiers are constructed based on the training set data. The system is trained with sample intrusions. These form the basis of identifying the pattern of user queries. The attribute vector for each of these intrusions is stored in a feature database. Two classes, namely, affected class and normal class, are considered here. All intrusions recorded in the database are classified into the same affected class. The instances of similar class are grouped into a single category. Two CA need to be built, where each CA is

trained to identify its class by estimating the optimal hyperplane for each CA.

2. Testing phase: In this phase, the attributes of the instance are used to form a query. The attribute vector is calculated and given as input to the pool of trained CA to identify the class of the instance and to take necessary steps.

8.4.3 Performance Comparison

After performance analysis, it has been seen that EEHIDS results in much better utilization of the available energy than existing SPAID. Figure 8.5 represents the comparative analysis between EEHIDS and existing SPAID. It has been noticed that the percentage of available energy is higher with an increase in node density in EEHIDS than in SPAID.

In EEHIDS, partitioning larger networks to clusters and manipulating energy levels and thresholds provides a more energy-optimal solution than that of SPAID, which considers the entire network. Also, SPAID was considered only for minimum mobility of the networks. EEHIDS with tree-based clusters can be efficient in the case of dynamic wireless networks.

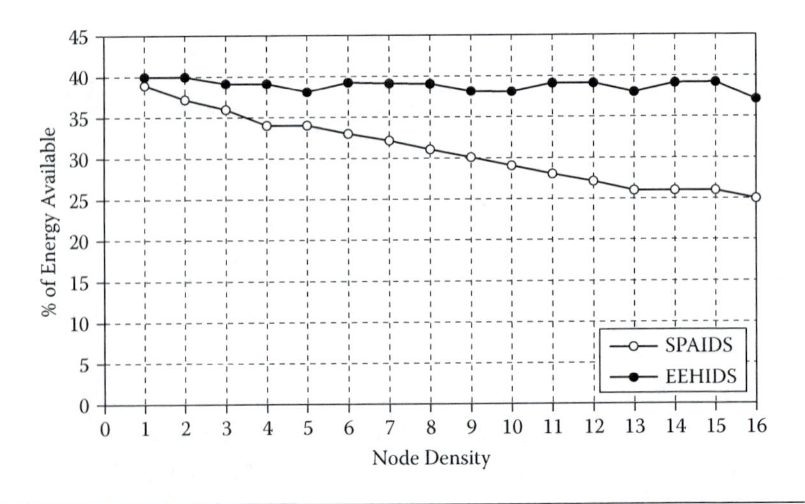

Figure 8.5 Comparison between EEHIDS and SPAID.

8.5 Summary

In this chapter, a brief overview of the future trends in WAN security has been provided. This chapter mainly focuses on the requirement of secure cloud services and smart grid architecture to the limited-resource WAN environment. It also presents a security architecture of cloud in WAN. Finally, it gives an idea about energy-efficient intrusion detection architecture in WAN using cellular automata. It represents the efficient functioning of the EEHIDS algorithm to avoid complete exhaustion of a node or network. The results show that the EEHIDS algorithm gives good results on any type of wireless ad-hoc network. As is evident from the energy utilization performance evaluation, the EEHIDS algorithm proves to be scalable, and even more efficient as the size of the wireless ad-hoc network increases.

References

1. K. Hamlen, M. Kantarcioglu, L. Khan, and B. Thuraisingham, Security Issues for Cloud Computing, *International Journal of Information Security and Privacy*, 4(2), 39–51, 2010.
2. D. Huang, X. Zhang, M. Kang, and J. Luo, MobiCloud: Building Secure Cloud Framework for Mobile Computing and Communication, at 2010 Fifth IEEE International Symposium on Service Oriented System Engineering.
3. J. Gao, J. Wang, and B. Wang, Research on Communication Network Architecture of Smart Grid, in *2012 International Conference on Future Electrical Power and Energy Systems Lecture Notes in Information Technology*, vol. 9.
4. T. Srinivasan, J. Seshadri, J.B. Siddharth Jonathan, and A. Chandrasekhar, *A System for Power-Aware Agent-Based Intrusion Detection (SPAID) in Wireless Ad hoc Networks*, Springer-Verlag, Berlin, 2005.
5. O. Kachirski and R. Guha, Efficient Intrusion Detection Using Multiple Sensors in Wireless Ad Hoc Networks, at 36th Annual Hawaii International 2003.
6. L. Zhou and Z.J. Haas, Securing Ad hoc Networks, *IEEE Networks Special Issue on Network Security*, November 1999.
7. P. Kiran Sree, Ramesh Babu, J.V.R. Murty, R. Ramachandran, and N.S.S.S.N. Usha Devi, Power-Aware Hybrid Intrusion Detection System (PHIDS) Using Cellular Automata in Wireless Ad-Hoc Networks, *WSEAS Transactions on Computers*, 7(11) November, 2008, pp. 1848–1874.

8. A. Abduvaliyev, S. Lee, and Y.K. Lee, Energy Efficient Hybrid Intrusion Detection System for Wireless Sensor Networks, at Proceedings of 2010 International Conference on Electronics and Information Engineering (ICEIE 2010).
9. A.M. Safwat, H.S. Hassanein, and H.T. Mouftah, Power-Aware Wireless Mobile Ad hoc Networks, in *Handbook of Ad hoc Wireless Networks*, CRC Press, Boca Raton, FL, December 2002.
10. D. Dasgupta and H. Brian, Mobile Security Agents for Network Traffic Analysis, in *Proceedings of DARPA Information Survivability Conference and Exposition II, DISCEX'01*, 2001, vol. 2, pp. 332–340.
11. G. Helmer, J. Wong, V. Honavar, and L. Miller, *Lightweight Agents for Intrusion Detection*, Technical Report, Department of Computer Science, Iowa State University, Ames, 2000.
12. M.C. Bernardes and E.S. Moreira, Implementation of an Intrusion Detection System Based on Mobile Agents, in *Proceedings of the International Symposium on Software Engineering for Parallel and Distributed Systems*, 2000, pp. 158–164.
13. Tao, J., L. Ji-ren, and Q. Yang, The Research on Dynamic Self-Adaptive Network Security Model Based on Mobile Agent, at Proceedings of the 36th International Conference on Technology of Object-Oriented Languages and Systems 2000.
14. S. Misra, P. Venkata Krishna, and K.I. Abraham, Energy Efficient Learning Solution for Intrusion Detection in Wireless Sensor Networks, at Proceedings of the 2nd International Conference on Communication Systems and Networks COMSNETS 2010.
15. Y. Zhang and W. Lee, Intrusion Detection in Wireless Ad-Hoc Networks, in *Proceedings of the 6th Annual International Conference on Mobile Computing and Networking, MobiCom*, 2000, pp. 275–283.
16. L.M. Feeney and M. Nilsson, Investigating the Energy Consumption of a Wireless Network Interface in an Ad hoc Networking Environment, at Proceedings of IEEE INFOCOM, 2001.
17. Hilda Larina Ragragio and Marian Radu, The Cloud or the Mist? at Virus Bulletin Conference, September 2009, http://scholar.googleusercontent.com/scholar?q=cache:dyhUyGony9IJ:scholar.google.com/+preventing+SQL+injection+attacks+in+the+cloud&hl=en&as_sdt=0,5.
18. P. Vogt, F. Nentwich, N. Jovanovic, E. Kirda, C. Kruegel, and G. Vigna, Cross-Site Scripting Prevention with Dynamic Data Tainting and Static Analysis, at Proceedings of the Network and Distributed System Security Symposium (NDSS'07), February 2007.
19. Tim Mather, Subra Kumaraswamy, and Shahed Latif, *Cloud Security and Privacy: An Enterprise Edition on Risks and Compliance (Theory in Practice)*, O'Reilly Media, Sebastopol, CA, 2009, http://oreilly.com/catalog/9780596802776.
20. Zouheir Trabelsi, Hamza Rahmani, Kamel Kaouech, and Mounir Frikha, Malicious Sniffing System Detection Platform, in *Proceedings of the 2004 International Symposium on Applications and the Internet (SAINT'04)*, 2004, pp. 201–207.

21. Kellep Charles, Google's Gmail Hacked by China Again, SecurityOrb, The Information Security Knowledge Base Website, June 2, 2011, http://securityorb.com/2011/06/googles-gmail-hacked-by-china-again/.
22. Rodrigo N. Calheiros, Rajiv Ranjan, Anton Beloglazov, Cesar A.F. De Rose, and Rajkumar Buyya, CloudSim: A Toolkit for Modeling and Simulation of Cloud Computing Environments and Evaluation of Resource Provisioning Algorithms, *Software: Practice and Experience (SPE)*, 41(1), 23–50, 2011.
23. John E. Dunn, Spammers Break Hotmail's CAPTCHA Yet Again, *Techworld*, February 16, 2009, http://news.techworld.com/security/110908/spammers-break-hotmails-captcha-yet-again/.
24. Albert B. Jeng, Chien Chen Tseng, and Der-Feng Tseng, Jiunn-Chin Wang, A Study of CAPTCHA and Its Application to User Authentication, at Proceedings of 2nd International Conference on Computational Collective Intelligence: Technologies and Applications, 2010.
25. Peter Mell and Timothy Grance, *The NIST Definition of Cloud Computing*, U.S. National Institute of Standards and Technology ITL Technical Report, January 2011, http://docs.ismgcorp.com/files/external/Draft-SP-800–145_cloud-definition.pdf.
26. C. Borovick, *IDC Analyst Connection*, June 2011, http://www.bluecoat.com/sites/default/files/analyst-reports/documents/IDC%20 Analyst%20 Connection%20NGWO%20.pdf.
27. R. Chaki and N. Chaki, IDSX: A Cluster Based Collaborative Intrusion Detection Algorithm for Mobile Ad-Hoc Network, in *Proceedings of the 6th International Conference on Computer Information Systems and Industrial Management Applications (CISIM'07)*, Minneapolis, MN, 2007, p. 179.
28. P. Techateerawat and A. Jennings, Energy Efficiency of Intrusion Detection Systems in Wireless Sensor Networks, at IEEE WIC 2006.
29. T. Bhattasali and R. Chaki, A Survey of Recent Intrusion Detection Systems in Wireless Sensor Network, in *Proceedings of the Fourth International Conference on Network Security and Applications*, 2011, pp. 268–280.
30. T. Bhattasali and R. Chaki, Lightweight Hierarchical Model for HWSNET, *International Journal of Advanced Smart Sensor Network Systems (IJASSN)*, 1(2), 17–32, 2011, DOI: 10.5121/ijassn.2011. 1202.
31. A. Fathi Navid and A.B. Aghababa, Irregular Cellular Learning Automata-Based Method for Intrusion Detection in Mobile Ad hoc Networks, *IndJST*, 6(3), 2013, http://test.fitce.org/congress/2012/papers/3.5-Irregular%20Cellular%20Learning.pdf.
32. J. Kari, Theory of Cellular Automata: A Survey, *Journal on Theoretical Computer Science*, 334(1–3), 3–33, 2005.
33. U.S. Agency for International Development, USAID India, *The Smart Grid Vision for India's Power Sector*, White Paper, March 2010.
34. Department of Energy Office of Electricity Delivery and Energy Reliability, *Study of Security Attributes of Smart Grid—Current Cyber Security Issues*, Technical Report, National SCADA Test Bed, April 2009.

35. Alvaro A. Cadenas and Ricardo Moreno, Cyber-Physical Security for Smart Grid Systems, at NIST Cyber Security for Cyber-Physical Systems Workshop, April 23–24, 2012.
36. P. Parikh, M.G. Kanabar, and T.S. Sidhu, Opportunities and Challenges of Wireless Communication Technologies for Smart Grid Applications, at IEEE PES General Meeting, Minneapolis, MN, July 25–29, 2010.
37. National Institute of Standards and Technology (NIST), *NISTIR 7628: Smart Grid Cyber Security: Smart Grid Cyber Security Strategy, Architecture, and High Level Requirements*, vol. 1, Smart Grid Interoperability Panel, Cyber Security Working Group, Washington, DC, August 2010.
38. Richard J. Campbell, *The Smart Grid and Cyber Security—Regulatory Policy and Issues*, CRS Report for Congress, R41886, June 15, 2011.
39. National Institute of Standards (NIST), *NISTIR 7628: Smart Grid Cyber Security: Privacy and the Smart Grid*, vol. 2, Smart Grid Interoperability Panel, Cyber Security Working Group, Washington, DC, August 2010.
40. T. Baumeister, *Literature Review on Smart Grid Cyber Security*, Technical Report, Collaborative Software Development Laboratory, Department of Information and Computer Sciences, University of Hawaii, December 2010.
41. Zhong Fan, Georgios Kalogridis, Costas Efthymiou, Mahesh Sooriyabandara, Mutsumu Serizawa, and Joe McGeehan, The New Frontier of Communications Research: Smart Grid and Smart Metering, in *International Conference on Energy-Efficient Computing and Networking*, Passau, Germany, April 13–15, 2010, pp. 115–118.
42. M.G. Rosenfield, The Smart Grid and Key Research Technical Challenges, at Symposium on VLSI Technology (VLSIT), June 15–17, 2010.
43. P. McDaniel and S. McLaughlin, Security and Privacy Challenges in the Smart Grid, *IEEE Security and Privacy*, 7(3), 75–77, 2009.
44. Mihui Kim, A Survey on Guaranteeing Availability in Smart Grid Communications, *Advanced Communication Technology (ICACT)*, 314–317, 2012.
45. Eric D. Knapp and James Broad, eds., How Industrial Networks Operate, in *Industrial Network Security: Securing Critical Infrastructure Networks for Smart Grid, Scada, and Other Industrial Control Systems*, chap. 5, Syngress Publication, Boston, MA, 2011.
46. Gilbert N. Sorebo and Michael C. Echols, *Smart Grid Security: An End-to-End View of Security in the New Electrical Grid*, CRC Press, Taylor & Francis Group, Boca Raton, FL, 2012.

Index

D

G

H

I

K

L

M

T

Z